WHAT EVERY ENGINEER SHOULD KNOW ABOUT ROBOTS

WHAT EVERY ENGINEER SHOULD KNOW
A Series

Editor

William H. Middendorf

Department of Electrical and Computer Engineering
University of Cincinnati
Cincinnati, Ohio

Other volumes in preparation

WHAT EVERY ENGINEER SHOULD KNOW ABOUT ROBOTS

Maurice I. Zeldman

President
Emzee Associates
Pittsburgh, Pennsylvania

MARCEL DEKKER, INC.   New York and Basel

Zeldman, Maurice I.
    What every engineer should know about robots.

    (What every engineer should know ; v. 11)
    Bibliography: p.
    Includes index.
    1. Robotics.   I. Title.   II. Series.
TJ211.Z45   1984   629.8'92   83-21064
ISBN 0-8247-7123-0

MARCEL DEKKER, INC.

270 Madison Avenue, New York, New York 10016

Current printing (last digit):

10 9 8 7 6 5 4 3 2 1

PRINTED IN THE UNITED STATES OF AMERICA

Preface

This book is aimed at engineers and manufacturing managers. Its purpose is to give sufficient data and background so that the technologists can help in deciding about a corporate investment in robotics. This book is a summary of the technology and economics of robotics today. A knowledge of this technology and its applications is the key to massive productivity improvements in special areas of the factory. These areas include heavy, difficult, boring, or repetitive tasks, as well as those when quality and consistency are crucial. In these particular areas, the economic returns on robotics can be immense.

In today's world, competition is international. Local problems of work force and of labor versus management can give either group the feeling of winning a battle; but without productive output, we, the country, may all lose the international economic "war." This book will help manufacturing and engineering managers evaluate the application of robots for their specific industry's needs. With knowledge of robotics, engineers can make recommendations about today's "state-of-the-art," and its applicability to the company's business. The field of robotics is growing rapidly today, and an investment in this technology might be premature for some companies. Only through a knowledge of the field can recommendations be made intelligently.

We begin "What Every Engineer Should Know About Robots" by defining the background, history, and trends of robots in the U.S. Next we focus on the domestic robot market and on Japan's role in robotic history. To bring the engineers and manufacturing managers to a working knowledge of the "state-of-the-art," the book defines basic robot terminology, and then leads these technologies through the key issues of Artificial Intelligence (AI). The development of AI is currently concerned with sensory robots, but the final impact will be the factory of the future.

A portion of the book is dedicated to productivity improvements resulting from robots. We focus on where robots pay off today. To aid in the selection of a robot system, we perform a Robot Cost Justification Study. Generic data are tabulated for systems available in the U.S. today. A separate insert tabulates up-to-date information on robots available in the U.S. One of our hopes in writing this book is that we will, in our way, contribute to the reindustrialization of America.

I would like to acknowledge the assistance of my family in the preparation of this book—specifically my son Jeffrey for acting as a creative editor, and my wife Phyllis for her assistance in pulling the manuscript together. I also would like to acknowledge all of the robot manufacturers for the agreeable way in which they supplied information and photographs.

Maurice I. Zeldman

Acknowledgments

We would like to thank the following robot manufacturers for use of their photographs: Figures 11, 12, and 21. Courtesy of Unimation, Inc., a Westinghouse Company. Figures 13 and 14. Courtesy of Cincinnati Milicron. Figure 15. Courtesy of Prab Robots, Inc. Figure 16. Courtesy of ASEA, Inc. Figure 17. Courtesy of The Devilbiss Company, a Division of Champion Spark Plug. Figure 18. Courtesy of IBM Corporation. Figure 19. Courtesy of General Electric Company. Figure 20. Courtesy of Cybotech, Inc. Figure 22. Courtesy of Mobot Corporation. Figure 23. Courtesy of Seiko Instruments USA, Inc.

List of Tables

List of Figures

Contents

About the Author

MAURICE I. ZELDMAN is President of EMZEE Associates, Inc., Pittsburgh, Pennsylvania, a technical and management consulting group which he founded in 1973. Mr. Zeldman has consulted for several robot manufacturers on technological and service problem areas, and conducts professional training programs on robotics and project management for technical and general managers. A pioneer in the field, he worked for AMF as the Chief Engineer of Versatran, and as the Manager of Automation where he developed a low cost Versatran Robot system. Subsequently, Mr. Zeldman served as a Corporate Director of Technical Development for the Industrial Divisions of Rockwell International, where he planned, built, staffed, and operated an R & D Division for the development of new products and processes. Mr.

Zeldman received the B.M.E. degree (1955) from the College of the City of New York and the M.S. degree (1963) in Electronics and Mathematics from Adelphi University, Garden City, New York. He frequently lectures for the American Management Association, and is a licensed professional engineer.

WHAT EVERY ENGINEER SHOULD KNOW ABOUT ROBOTS

Robot History: Background

Machines which move under their own power have fascinated humankind since the beginnings of recorded history. One of Aristophanes' plays is on record as having employed entirely automatic puppets, which moved by means of weights and pulleys. Wind-up dolls have been part of European culture for hundreds of years.

But machines which can be "instructed" to perform simple human tasks are of recent vintage; and their history is both interesting and informative in the sense of how to — or how *not* to — use and develop industrial robots. Today's so-called "next generation" of experimental robots are being developed to cope with uncertainties. In essence, machines are being created which think, and, in the most primitive sense, feel and see. Robot development has so far branched into (on one leg) exploitation of available technology for short-term goals, and (on the other) scientific study of "artificial intelligence." This branching will, in the foreseeable future, come together.

TABLE I

ROBOT HISTORY

1922 · CZECH AUTHOR KAREL CAPEK'S DRAMATIZATION ROSSUM'S UNIVERSAL ROBOTS ("R.U.R.") COINED THE WORD ROBOT.

1939 · ISAAC ASIMOV WRITES A SERIES OF STORIES ABOUT ROBOTS. THESE CULMINATE IN A BOOK "I ROBOT" OF NINE SHORT STORIES PUBLISHED IN 1950. ASIMOV'S "THREE LAWS OF ROBOTICS" WERE DEVELOPED IN 1942.

1954 · GEORGE DEVOL JR. APPLIES FOR A PATENT FOR A PROGRAMMABLE TRANSFER DEVICE.

1956 · DEVOL LICENSES MANNING, MAXWELL AND MORE. A PRELIMINARY PRODUCT SPECIFICATION IS OUTLINED FOR THE UNIMATE ROBOT.

1958 · AMF INITIATES AN R&D PROJECT FOR A VERSATRAN (VERSATILE TRANSFER) ROBOT.
     · CONDEC ENTERS A LICENSE AGREEMENT WITH DEVOL.
     · UNIMATE PREPARES A DETAILED PERFORMANCE SPEC.

TABLE I (CONTINUED)

1962 · PULLMAN INC. AND CONDEC ENTER AN AGREEMENT TO FORM A NEW VENTURE, UNIMATION INCORPORATED. PULLMAN OBTAINS CONTROLLING INTEREST FOR $3 MILLION. CONDEC EXCHANGES PATENTS AND KNOWLEDGE FOR MONEY AND A 44% OWNERSHIP OPTION.

· AMF INTRODUCES MODEL #102, A CONTINUOUS PATH TRANSFER DEVICE, AND MODEL #212, A POINT TO POINT TRANSFER DEVICE.

· UNIMATE PREPARES MARK I PROTOTYPE FOR FIELD TESTING.

1965 · AMF LICENSES 2 EUROPEAN MANUFACTURERS.

· AMF STARTS DEVELOPMENT OF A LOW COST VERSATRAN SOLID STATE ROBOT SYSTEM.

· CONDEC REPURCHASES UNIMATE FROM PULLMAN FOR APPROXIMATELY $4 MIL-LION.

1966 · UNIMATE AND VERSATRAN BOTH GAIN FIELD EXPERIENCE. THEIR EXPERIENCE, STARTING IN 1962, CULMINATES IN IMPROVED PERFORMANCE. IT IS BASED ON AP-PLICATION EXPERIENCE IN INDUSTRIAL ENVIRONMENTS.

· VERSATRAN INTRODUCES MODEL 301, A POINT TO POINT SOLID STATE ROBOT.

1967 · UNIMATE INTRODUCES MARK II, 2000. THEIR CUMULATIVE EXPERIENCE WITH MARK II, 1900, IS BASED ON 70 INDUSTRIAL INSTALLATIONS.
· THE FIRST ROBOT IS IMPORTED TO JAPAN. AN AMF VERSATRAN IS USED ON A PAINT SPRAYING APPLICATION.

1968 · AMF VERSATRAN INTRODUCES MODEL 302, A POINT TO POINT ROBOT.
· PRAB ENTERS THE ROBOT MARKET WITH A MEDIUM TECHNOLOGY ROBOT.
· JAPAN DEVELOPS 2 SERVO ROBOTS WITH VERSATRAN MOTIONS.
· THE FIRST UNIMATE IS IMPORTED BY JAPAN'S KAWASAKI HEAVY INDUSTRIES.

1969 · THE FIRST UNIMATE IS INSTALLED IN JAPAN'S NISSAN AUTOMOBILE FACTORY.
· THE U.S. DEFENSE ADVANCE RESEARCH PROJECT AGENCY (DARPA) FUNDS ARTIFI-

CIAL INTELLIGENCE RESEARCH AT 5 UNIVERSITIES.
· THE UNIVERSITY OF EDINBURGH STARTS A "HAND-EYE" PROGRAM FOR ASSEM-
BLY ROBOTS.

1970 · UNIMATE INTRODUCES THE FIRST 6 DEGREE OF FREEDOM INDUSTRIAL ROBOT.
· AMF DEMONSTRATES ITS PRODUCTION MODEL 401 IMPROVED CONTINUOUS PATH INDUSTRIAL ROBOT.

TABLE I (CONTINUED)

- ITT PLANS A STRATEGIC ENTRY INTO THE ROBOT MARKET. THEY ARE INTERESTED IN THE LOW TECHNOLOGY PICK AND PLACE SEGMENT OF THE MARKET.

- AMF JAPAN LTD. IS FORMED. AT THIS TIME, 26 VERSATRANS ARE INSTALLED AND OPERATING IN JAPANESE FACTORIES.

1971 · THE JAPAN INDUSTRIAL ROBOT ASSOCIATION IS FORMED (JIRA).

1972 · ROBOTICS INCORPORATED ENTERS THE MARKET.

· IBM IS EXPERIMENTING WITH A RECTANGULAR COORDINATE SYSTEM ROBOT FOR INTERNAL USE. THE IBM CONCEPT PERMITS ANY LENGTH, HEIGHT, OR WIDTH. ROBOTS WILL BE USED BY IBM FOR INTERNAL MANUFACTURING ASSEMBLY ONLY.

· SUNSTRAND INTRODUCES AN ANTHROPOMORPHIC (HUMAN LIKE MOTIONS) ROBOT OF ADVANCED TECHNOLOGY. THIS SOPHISTICATED SERVO SYSTEM IS THE "STATE OF THE ART" IN ROBOTS FOR THIS TIME PERIOD. AFTER SEVERAL YEARS OF DIS-

APPOINTING SALES, SUNSTRAND WITHDRAWS FROM THE MARKET. THEY OFFER LICENSE AGREEMENTS.

· KANEMATSU GOSHO ENTERS THE INDUSTRIAL ROBOT MARKET IN JAPAN.

6

- ELECTROLUX IN SWEDEN ENTERS THE INDUSTRIAL ROBOT MARKET AFTER USING ROBOTS WHICH THEY DEVELOPED FOR INTERNAL MANUFACTURING USE OVER A 4-YEAR PERIOD.
- THE JAPANESE GOVERNMENT FUNDS ADVANCED ROBOTIC RESEARCH PROGRAMS. UNLIKE DARPA, THE JAPANESE ARE FOCUSING ON SOLUTIONS TO INDUSTRIAL PROBLEMS. THEY DO THIS BY FUNDING PRIVATE INDUSTRY R&D EFFORTS. (THIS CONTRASTS WITH U.S. FUNDING AT UNIVERSITIES.) THEY WILL SPEND $100 MILLION OVER THE NEXT 5 YEARS IN APPLIED ROBOT RESEARCH.

1973
- HAYES-DANA AND THE CANADIAN GOVERNMENT BUILD THREE PROTOTYPE INDUSTRIAL ROBOTS.
- WARNECKE AND SCHRAFFT OF STUTTGART UNIVERSITY CATALOGUE ALL OF THE ROBOT MANUFACTURERS IN THE WORLD IN "INDUSTRIE ROBOTER". AT THIS TIME, 71 FIRMS ARE SPECIFIED. MANY COMPANIES GET OUT OF THE BUSINESS WHEN THEY DO NOT FIND SHORT TERM PROFITS.

1974
- ASEA DEVELOPS AN ALL ELECTRIC INDUSTRIAL ROBOT. THEY OFFER THREE AND FIVE AXIS ANTHROPOMORPHIC UNITS.

TABLE I (CONTINUED)

1975 · A ROBOT SOCIETY IS FINALLY FORMED IN THE U.S. IT IS CALLED RIA (ROBOT IN-STITUTE OF AMERICA).

· UNIMATE HAS ITS FIRST PROFITABLE YEAR! MANY OF THE ORIGINAL ROBOT COM-PANIES ARE OUT OF BUSINESS.

1976 · CINCINNATI MILACRON ENTERS THE INDUSTRIAL ROBOT MARKET. THEY OFFER AN ANTHROPOMORPHIC SIX AXIS INDUSTRIAL ROBOT OF SOPHISTICATED TECH-NOLOGY.

· VW DISPLAYS A ROBOT SYSTEM AT THE HANOVER FAIR IN GERMANY.

1977 · THE BRITISH ROBOT ASSOCIATION IS FORMED (BRA).

1978 · GENERAL MOTORS FILLS ITS PROGRAMMABLE UNIVERSAL MACHINE FOR ASSEM-BLY (PUMA) CONCEPT WITH A CONTRACT FOR UNIMATE MANUFACTURE. UNIMATE STARTS TO BUILD PUMA ARMS. GM'S STUDY SHOWS THAT 95% OF ALL AUTO-MOTIVE PARTS WEIGH THREE POUNDS OR LESS. PUMA IS AN ANTHROPOMOR-PHIC ELECTRIC DC SERVO DRIVE ROBOT WITH 5 OR 6 AXIS CAPABILITY, RANGING FROM 2 POUND PAYLOAD TO 5.5 POUNDS.

1979 • PRAB PURCHASES VERSATRAN. THIS PURCHASE COMBINES HIGH AND MEDIUM TECHNOLOGY ROBOT MANUFACTURERS.

1980 • THE SOCIETY OF MANUFACTURING ENGINEERS (SME) FORMS ROBOTICS INTERNATIONAL (RI). THE SOCIETY PUBLISHES A QUARTERLY MAGAZINE, "ROBOTICS TODAY".

• AMERICAN ASSOCIATION FOR ARTIFICIAL INTELLIGENCE (AAAI) IS FORMED. THE SOCIETY PUBLISHES AI MAGAZINE.

• AUTOMATIX INC. IS FORMED. THEY SPECIALIZE IN TURN-KEY ROBOT INSTALLATIONS WHICH COMBINE SENSORY, SOFTWARE, AND ROBOTS. THE COMPANY GOAL IS TO COMBINE CAM (COMPUTER AIDED MANUFACTURER) WITH ROBOTICS.

• CYBOTECH, A JOINT VENTURE BETWEEN RANSBURG AND RENAULT IS FORMED. THEY PRODUCE ALL SPECIES OF ROBOTS. FAMILIES INCLUDE CYLINDRICAL COORDINATE SYSTEMS, X, Y, Z AND ANTHROPOMORPHIC ROBOTS.

1981 • PLANET, ONE OF THE FOUNDING ROBOT MANUFACTURERS, CHANGES PLANET ROBOT DIVISION TO "ARMAX ROBOTICS INC." THEY OFFER A LINE OF ROBOTS FROM SIMPLE PICK AND PLACE DEVICES, THROUGH SERVO SENSORY ROBOTS.

9

TABLE I (CONTINUED)

1981 • GENERAL ELECTRIC SIGNS A NON-EXCLUSIVE AGREEMENT WITH DEA ROBOT OF ITALY TO MANUFACTURE AND SELL INDUSTRIAL ROBOTS IN THE US. AT ONE TIME, GE MANUFACTURED AN IMPRESSIVE MANIPULATOR CALLED MAN-MATE. THIS AN-THROPOMORPHIC, SENSORY MANIPULATOR COULD HAVE BEEN EASILY CONVERT-ED TO AN INDUSTRIAL ROBOT SYSTEM. GE INTENDS TO LINK THEIR CALMA CAD SYSTEMS, ROBOTS AND OTHER GE AUTOMATION INTO AUTOMATED MANUFACTUR-ING PROCESSES.

1982 • GE SIGNS AGREEMENTS WITH VW AND HITACHI FOR SALE OF THEIR ROBOTS IN THE U.S. THESE AGREEMENTS GIVE GE A LARGE VARIETY OF ROBOTS WHICH RANGE FROM SMALL SIZE ASSEMBLY TYPE UNITS, THROUGH INTERMEDIATE SIZES AND INTO THE LARGE HEAVY DUTY SIZES. A VARIETY OF SENSORY FEED-BACK AND GRIPPER OPTIONS ARE OFFERED WITH THE ALLEGRO ASSEMBLY ROBOT WHICH IS CAPABLE OF USING ONE TO FOUR ARMS. GE ALSO OFFERS OTHER UNITS THAT ARE USED IN PROCESS, PAINT SPRAY, AND TRANSFER APPLICATIONS.

• IBM ENTERS THE ROBOT FIELD WITH ASSEMBLY ROBOTS THAT THEY HAVE USED INTERNALLY FOR ASSEMBLY OF TYPEWRITERS AND COMPUTERS. THEIR MARKET THRUST IS AIMED AT SMALL PRECISE ASSEMBLY ROBOTS. THESE UNITS USE A

10

SPECIAL ROBOTIC COMPUTER LANGUAGE CALLED "AML" (AN ASSEMBLY LAN-
GUAGE). TWO ROBOTS ARE OFFERED, ONE HYDRAULIC CARTESIAN COORDINATE
SIX DEGREE OF FREEDOM SYSTEM CALLED RSI, AND ONE HORIZONTAL ANTHRO-
POMORPHIC FOUR DEGREE OF FREEDOM SYSTEM CALLED 7535. BOTH ARE OP-
ERATED FROM A PERSONAL COMPUTER. EACH UNIT CAN BE EQUIPPED WITH SEN-
SORY FEEDBACK INFORMATION, TACTILE, INFRARED, AND SPECIAL FINGERS.

- WESTINGHOUSE ENTERS THE MARKET WITH A VARIETY OF ROBOTS THAT RANGE
IN SIZE FROM THE 1000 AND 2000 SERIES, SMALL ANTHROPOMORPHIC ASSEMBLY
ROBOTS, TO THE 4000 ANTHROPOMORPHIC WELDING ROBOT, WITH A 5000 ONE
OR TWO ARM CARTESIAN COORDINAL ASSEMBLY ROBOT, TO A 7000 SERIES
MASSIVE HEAVY DUTY CARTESIAN COORDINATE WELDING ROBOT. THEY ALSO OF-
FER TABLE WORKPIECE POSITIONERS AND A VARIETY OF SERVICES.

- CYBOTECH EXPANDS ITS LINE TO INCLUDE AN ELECTRIC ROBOT.

- CINCINNATI MILACRON IS PRODUCING TWO NEW ALL ELECTRIC ANTHROPOMOR-
PHIC ROBOTS. THE NEW T³746 AND T³726 ROBOTS BOTH USE A NEW "3 ROLL
WRIST". THE WRIST IS A COMPACT 3 DEGREE OF FREEDOM SERVO UNIT. COM-
BINED WITH T³ HYDRAULIC ROBOTS, MILACRON OFFERS A WIDE RANGE OF ENVI-
RONMENTAL AND LOAD CONFIGURATIONS.

- IN 1982 A LARGE GROUP OF NEW MANUFACTURERS APPEAR ON THE SCENE.

TABLE I (CONTINUED)

EVERY MANUFACTURER IS EXPANDING THEIR LINE. SMALL ROBOT MANUFAC-
TURERS START TO MAKE LARGER UNITS, AND VICE VERSA. HYDRAULIC MANUFAC-
TURERS BEGIN TO MAKE ELECTRIC UNITS AND VICE VERSA. THE DOMESTIC
ROBOT INDUSTRY IS ON A 36% GROWTH CURVE. THE INDUSTRY IS DIGGING IN
FOR A MARKET SHARE BATTLE. UNIMATE, THE INDUSTRY'S OLDEST ROBOT COM-
PANY, FACES STIFF COMPETITION. JAPANESE ROBOT MANUFACTURERS ARE TAR-
GETING U.S. MARKETS. AMERICAN INDUSTRIES ARE LOOKING TOWARD ROBOTS
TO SOLVE PRODUCTIVITY AND COST COMPETITION PROBLEMS.

· GENERAL MOTORS AND FANUC (A SPIN OFF OF FUJITSU WHICH CURRENTLY
HOLDS 50% OF THE WORLD MARKET FOR NUMERICAL CONTROL MACHINES) HAVE
AGREED TO A JOINT VENTURE IN THE UNITED STATES TO MANUFACTURE FANUC
DESIGNED MACHINE TOOL LOADING AND ASSEMBLY ROBOTS. GM CURRENTLY
BUILDS A PAINT SPRAY ROBOT WHICH WILL ALSO BE SOLD IN THE UNITED
STATES.

· WESTINGHOUSE ACQUIRED UNIMATE IN A SURPRISE MOVE IN DECEMBER OF 1982.
THIS MOVE WILL OPEN NEW MARKETS FOR BOTH COMPANIES. IT GIVES WESTING-
HOUSE HYDRAULIC CAPACITY, AND UNIMATE NEW ELECTRIC ROBOTS IN BOTH
THE LARGE CAPACITY AND SMALL ASSEMBLY AREAS.

Domestic Market for Robots 1980–1982

In 1980, the total domestic market was estimated to be $92 million. In that year, more than two-thirds of the U.S. market was shared by two manufacturers—Cincinnati Milacron, with a 37.5% share, and Unimate, with a 38.5% share. The less than a third remaining market was shared by Prab/Versatran, Devilbiss and ASEA, chiefly, with all others producing a mere 5.8% of the total. A shift in this market segmentation is beginning to occur. In 1981, the All Other segment expanded with new companies, one of which is GE. In 1982, IBM and Westinghouse entered the market while GE expanded License Agreements with VW and Hitachi, and finally an American auto manufacturer, GM, is entering the market in a joint agreement with Fanuc.

TRENDS IN THE
ROBOT MARKET

By comparing 1980 and '81 Market Share in Figures 1 and 2, we can see that the robot market will change radically in the near future, as more and more manufacturers join the expanding bandwagon. Naturally the new manufacturers will deplete some of the market share of the major domestic robot manufacturers—Cincinnati Milacron, Unimate, ASEA, Prab/Versatran and Devilbiss. To try to hold their share, they must try to grow rapidly. IBM, GE, Cybotech and Westinghouse for their part, will try to overtake them by 1985.

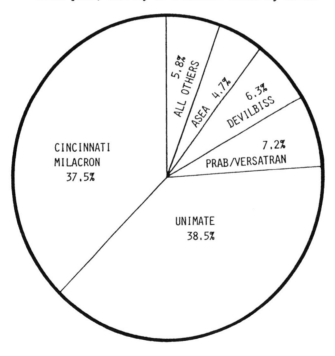

Figure 1. Estimated Share 1980 Domestic Industrial Robot Market. Total Domestic Market is $92 Million.

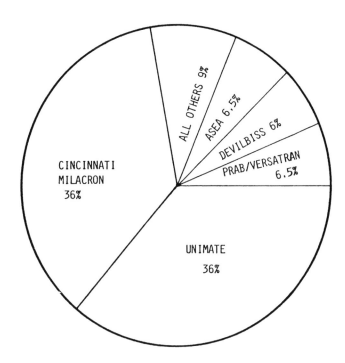

Figure 2a. Estimated Share 1981 Domestic Industrial Robot Market. Total Domestic Market is $134 Million.

Westinghouse, in a surprise move, acquired Unimate in December, 1982. This linking will give Unimate access to the Westinghouse electric technology. It will give Westinghouse the Unimate hydraulic technology, and the oldest name in the robotic business. The combination could be the chemistry that both groups need to grow.

The domestic market share for 1982 outlined in Figure 3, shows that the robot market continued to grow through a poor business year. Although total robot sales increased, the robot makers' profits diminished. A large number of the robot manufacturers ran deficits. These deficits, due largely to in-

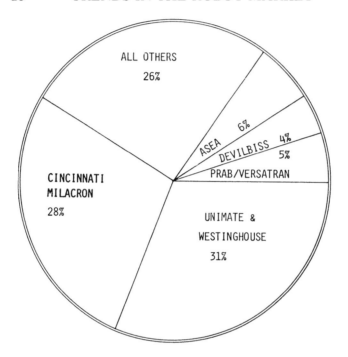

Figure 2b. Estimated Share 1982 Domestic Industrial Robot Market. Total Domestic Market is $180 Million.

creased competition, will most likely squeeze smaller companies out of the market. Deficits continue to accelerate acquisitions in 1983. This trend toward growth through acquisition and combination will continue until the market settles down.

The total market for 1982 is estimated at 180 million dollars. It has almost *doubled* in size in two years. For most of the older manufacturers the market remained flat or shared modest growth. The bulk of the market growth occurred in the segment marked other. Contributing to this 26% market share are companies that include IBM, GE, Cybotech, and Mobot. Some of these companies now have a market share greater than

ASEA or Prab/Versatran. Westinghouse and Unimate sales have combined in figures. This combination gives them a sales edge over Milacron for 1982.

ASEA has made its growth plans known. It intends to expand in Japan and compete with Japanese manufacturers for both Japanese and world markets. ASEA's market share has shifted in the U.S. dramatically in the year between 1980 and 1981 with company dedication to growth in this area.

Cybotech Corp., a joint venture company of the Ransburg Corporation of Indianapolis (51% owner) and Renault of France (49%) is rapidly building and selling its robots. They targeted an ambitious 4% share of the market by the end of 1982. Using Renault's experience in robot manufacture and the French auto manufacturer's robot application experience, Cybotech has entered the field with a "running start".

IBM has been using internally developed robots and a small Japanese robot mechanism for the manufacture of typewriters and computers over the past four years. They entered the market, in 1982, with experience in small assembly. Assembly is at the forefront of today's robot technology. Both the IBM developed hydraulic robot and the Japanese electric robot that they offer are operated through a "personal" computer. This computer makes use of a new Robot Language called AML (a manufacturing language). IBM's experience gives them leverage in the expanding businesses of computers and small mechanisms.

GE, through License Agreements with DEA of Italy, Hitachi of Japan and VW of Germany, is planning to "cover the market" with any and all mechanisms and arrangements needed to automate production.

The companies discussed and other new arrivals, shown in the tabulation of robot manufacturers, appended to this book, will challenge the current leaders. This does not imply that the current leading robot manufacturers will permit this challenge to go without resistance.

Cincinnati Milacron has developed two new all electric robots and a special wrist mechanism. This equipment was displayed at the 1982 Robots VI Conference and Exposition. Until 1982, Milacron sold hydraulic robots. They are now competing directly with ASEA and the other all electric robot manufacturers. Their reputation for service, coupled with their skill and production of numerically controlled machine tools, gives them a strong edge in the marketplace. Milacron will be in a good position to build the fully automated factories of the future.

Westinghouse has built a productivity center in Pittsburgh. Their intention is to expand robotics in the assembly and welding areas. They have formed strong ties with the Carnegie Mellon robotic group. Several robots and peripherals are being manufactured to provide "turn-key" assembly and manufacturing capability.

Unimate, the "Grand Old Robot Company", now a part of Westinghouse, has the most cumulative experience of any manufacturer. This experience combined with Westinghouse technology could be the needed push that Unimate requires to maintain or improve their market share. The new manufacturers are all attempting to carve out a specialized market niche for their offerings. Unimate and Milacron will be tougher competitors than they were before.

Prab Conveyors' acquisition of Versatran, the second oldest robot company, is beginning to pay off. This combination brings together the sophisticated Versatran and the simple Prab Robots. At this time the group lacks electric robots. Historically, Versatran's major problem was small sales volume. Growth of the overall industry will bring them the advantage of increased volume.

The business concept that Devilbiss/Trallfa has capitalized upon is specialization. A division of Champion Spark Plug, based in Norway, Devilbiss/Trallfa is the largest paint spray robot manufacturer in the world. Some of the new manufacturers are aiming at their market.

The early robot market in the U.S. grew on a haphazard basis. With increased business emphasis on making America more productive, the robot manufacturers that "win the day" will be the strategists and planners of new manufacturing facilities. These manufacturers recognize the robot as one of many pieces of equipment in an Artificially Intelligent Factory. Computer Aided Design (CAD) and Integrated Computer Aided Manufacture (ICAM) combined with robots and numerically controlled machines will provide the framework for the factory of the future.

The current business trend in the Robotics industry seems to be that manufacturers of robots alone will eventually be absorbed by companies who will offer the "Turn-Key" automated factories of the future.

The following tables are divided by robot *type*. Smallest in world population are programmable servo robots — those with a slave control system. Programmable non-servo robots are more widely used in the world markets (these employ no direct feedback). Pick and place robots, which many would call simple automation and not true robots are the most widely used outside the U.S. Clearly, the application of "thinking machines" or Artificial Intelligence is a science still coming of age.

It is interesting to observe that the U.S., while one-hundred-and-twenty-five times the size of Japan, employs fewer robots — even programmable servos — than Japan does. Of the servos, roughly two-thirds of those America uses are employed in the automobile plants, with a large minority used in the die-casting industry. By the way, that figure for Western Europe shown in Tables II, III and IV includes Poland. Even though the U.S.S.R. is not shown in the tables, our latest data indicates a total of just twenty-five servo robots are currently in use in the U.S.S.R.

Further examination of Table III shows that while the U.S. comes closest to Japan in its population of programmable servos, their pick-and-place population is nearly eighteen times greater than ours. Figures for pick-and-place robots in the U.S. may actually be higher than indicated, as some types designed-in-house for in-house use would not show up in our figures.

In Table IV the 1982 robot census shows that the U.S. is beginning to use more pick and place commercial robots and that the ratio between the U.S. and Japan has narrowed to approximately fourteen times. One of the major shifts that we will see in the U.S. robot population over the next decade will be in the pick and place robot population. These units will be used in conjunction with conveyor systems for changes of elevation, and for loading applications. The semi-intelligent conveyor system will be an output of this trend.

TABLE II
ROBOTICS: 1980 WORLD POPULATION OF INDUSTRIAL ROBOTS

COUNTRY	PROGRAMMABLE SERVO ROBOTS (POINT TO POINT AND CONTINUOUS PATH)	PROGRAMMABLE NON SERVO ROBOTS	MECHANICAL TRANSFER DEVICES (PICK AND PLACE ROBOTS)
JAPAN	3,000 (46%)	11,000 (84.6%)	33,000 (83.6%)
U.S.	2,155 (33%)	1,100 (8.5%)	1,075 (2.7%)
SWEDEN	570 (9%)	NA (NOT AVAILABLE)	NA
W. GERMANY	450 (7%)	400 (3.1%)	5,000 (12.6%)
WESTERN EUROPE*	335 (5%)	500 (3.8%)	410 (1.1%)
TOTAL NUMBER IN EACH CLASS TOTAL = 59,005	6,510 (100%)	13,000 (100%)	39,495 (100%)
PERCENT OF TOTAL ROBOTS	(11%)	(22%)	(67%)

22

TABLE III

ROBOTICS: 1981 WORLD POPULATION OF INDUSTRIAL ROBOTS

COUNTRY	PROGRAMMABLE SERVO ROBOTS (POINT TO POINT AND CONTINUOUS PATH)	PROGRAMMABLE NON SERVO ROBOTS	MECHANICAL TRANSFER DEVICES (PICK AND PLACE ROBOTS)
JAPAN	4,150 (46.6%)	14,850 (85%)	51,000 (82.5%)
U.S.	2,910 (32.7%)	1,480 (8.5%)	2,915 (4.7%)
SWEDEN	780 (8.8%)	NA (NOT AVAILABLE)	NA
W. GERMANY	550 (6.2%)	450 (2.6%)	7,000 (11.3%)
WESTERN EUROPE*	450 (5.0%)	680 (3.9%)	565 (0.9%)
TOTAL NUMBER IN EACH CLASS TOTAL = 88,120	8,900 (100%)	17,460 (100%)	61,760 (100%)
PERCENT OF TOTAL ROBOTS	(10%)	(20%)	(70%)

23

TABLE IV

ROBOTICS: 1982 WORLD POPULATION OF INDUSTRIAL ROBOTS

COUNTRY	PROGRAMMABLE SERVO ROBOTS (POINT TO POINT AND CONTINUOUS PATH)	PROGRAMMABLE NON SERVO ROBOTS	MECHANICAL TRANSFER DEVICES (PICK AND PLACE ROBOTS)
JAPAN	5,395 (47.1%)	19,450 (86.2%)	66,300 (82%)
U.S.	3,800 (33.2%)	1,780 (7.9%)	4,600 (5.8%)
SWEDEN	950 (8.3%)	NA (NOT AVAILABLE)	NA
W. GERMANY	770 (6.7%)	540 (2.4%)	9,100 (11.3%)
WESTERN EUROPE*	540 (4.7%)	780 (3.5%)	750 (0.9%)
TOTAL NUMBER IN EACH CLASS TOTAL= 114,815	11,455 (100%)	22,500 (100%)	80,810 (100%)
PERCENT OF TOTAL ROBOTS	(10%)	(20%)	(70%)

Japanese Robot History

The first robot imported into Japan was a Versatran. It arrived on December 5, 1967. The purchaser was Toyoda Automatic Loom Company, the parent of Toyota Automobile. It is interesting to observe what happened after that. Kawasaki Heavy Industries signed an agreement with Unimate in 1968, and Japan started to manufacture robots. A robot society was formed in 1971. In 1971 the Japanese government designated Industrial Robotics as one of the industries that qualified for special tax and financial provisions needed to promote their growing manufacturing technology. In 1978 a new law permitted any company that installed a robot to deduct an additional 12.5% in the first year in addition to normal depreciation. This is typical of Japanese governmental cooperation with its fledgling industries.

The history of robots in Japan is not unlike the history of NC machines and semi-conductors. The first steps in all of these fields, those of invention, research and development, were carried out in the U.S.A., often by independent companies, but occasionally with U.S. government R&D funds.

In all three development areas (robots, NC machines, and semi-conductors) the Japanese entered the field late. All of their work was developed using U.S. technology as a base. In each case, although the Japanese entered the field late, they worked hard to close their initial technological gap. As Japan became competent in the "technological art", their emphasis shifted to the area of their post war strength — production engineering and quality control. In NC and robotics Japan caught up to and passed the U.S.

Because Japan uses capital differently than we do, their industries advance beyond the U.S. level. Concentrating on the intelligent use of new capital, they begin to produce quality products at competitive costs. Japan gains in market share, and when they finally dominate the market, they become exporters of technology.

Presently Japan is ahead of the U.S. in robots, Numerical Control machines (FANUC has more than 50% of the world market) and in some limited sections of the semi-conductor industry.

The Japanese pattern of growth makes use of innovations that were developed in the U.S., but that were not necessarily fully funded by U.S. industry.

The U.S. has traditionally placed its emphasis on office automation, while the Japanese emphasis has been in precision manufacturing and production technology. Japan leads the world in development of the "manless" factory. Even though they are currently behind the U.S. in innovation and new developments, Japanese emphasis has caused industrial growth that will eventually attract innovators. Unless the U.S. pays attention to production automation, we will eventually lose the technical edge that has historically brought growth and prosperity. In the arena of automation, we are currently behind Japan. In solid state technology, the balance is beginning to shift in their favor. The use of capital may be the determining factor.

There are currently over 120 robot companies in Japan. Some of the larger companies are now targeting American markets. Companies like GE, IBM, and GM, to name a few, are selling Japanese robots domestically. This trend will continue. Domestic robot manufacturers, the inventors of the technology, will have to stay on the razor edge of technology to maintain market share. This decade will be decisive to the survival of the NC, CAD, and robots industries of America.

Robot Industry Future Projections

Comparing Tables II and III in the one-year span between 1980 and 1981, the population growth rate for all robots in the world was 49%. Data for each segment shows that servo robots had a 37% growth rate, programmable non-servo had a 34% growth rate with pick-and-place having a 70% growth rate. We believe that the pick-and-place simple mechanisms will become the largest growth segment of the domestic robot industry over the coming decade. Because they are the least expensive segment of the market, they will not have a big impact on the financial statistics of the total market. By the turn of the century, servo robots, due to high volume production, will probably drop in price to $10,000. At the lower price, they will be considerably affordable. Even small businesses will be using robots.

Considering overall market growth and the price drop for an average robot, we forecast the following future market. Data for the United States:

Year	Total Market in Constant Dollars
1985	$450 Million
1990	$ 2 Billion
1995	$ 8 Billion
2000	$ 20 Billion

By the turn of the century, the Japanese market could be two times the U.S. Domestic market. Many factors could work to change this projection—the socio-economic factors must not be ignored. Labor's role in automating America is essential to the success of the robot industry.

Basic Robot Terminology

Corporate decision makers and small business CEO's, in attempting to determine the impact of robotics on the growth of their businesses, are often confounded by robotic terms or jargon. In this section of the special report we will clarify basic robot terms, robot types, and special considerations. In subsequent sections, we will deal with specific suppliers of robots, their offerings, maintenance availability and productivity related concepts.

The basic terms in this section are defined in a chronological order. We start with the earliest robots, types, and terms, and work toward the most up-to-date technology.

SERVO SYSTEMS

Servo means "slave" and servo systems are just that: machines that move by command of a computer or control device. Servo systems are feedback systems. Originally developed in the 1920's for regulators and remote steering, servo systems were perfected during World War II for gun control. A servo system works to null an error, or command signal. Command signals fed into the input cause the system to drive a feedback transducer which is coupled to the output motion, in a direction that will balance or null the original input signal. The earliest robots and the more sophisticated systems of today, use multiple linked servo axes which can be designed to provide continuous motion in a limited space.

**POINT TO POINT
SYSTEMS**

Digital information is processed on a discreet basis. When single signals are fed into a servo robot, the system moves to that point. When another signal is fed into the system, the robot servo mechanism moves to follow that point. Continuous feeding of signals causes displacement from point to point. Depending upon the complexity of the computer software or point to point generating system, robots can be made to cover thousands of points in space, thereby approximating a continuous path, just as the quickly changing static images of a motion picture approximate the continuous motion of life. Although point to point systems have in the past been considered relatively limited, they have become more and more versatile as digital memory systems improve. Moreover, the state of the art today is such that point to point systems can be held to be as versatile as continuous path systems. A large number of digital points can be made to approach the smooth proportional motions of continuous path systems found in early robots with considerably greater reliability in factory environments. With today's technology, the distinction between point to point and continuous path is becoming archaic.

CONTINUOUS PATH SYSTEMS

Continuous path systems generate an ongoing flow of analogue or proportional signals into the servo mechanisms. With the proper response, the robot system moves along the signal path describing a smooth contour in space. Spacial contours are not generated by point to point. They are analogue signals (like the audio signals on magnetic tape) which can, with modern computer technology, be generated by calculation for each robot axis.* In the early robot, systems signals were produced by moving the robot arm through its task while in a teaching mode. Transducers (i.e., potentiometers) coupled to the robot's arm, produced these signals, which were then recorded on tape. Tape playback generated commands, or error signals were fed into the servo systems. This produced the continuous path spacial signals needed to repeat the original operator motions.

------

*Each possible motion that a robot can move along is called an axis.

SLEWING

Point to point systems can be made to look like continuous path systems by slewing from one point to another. Before the error signal has been nulled, the computer moves to a new point. Slewing permits the robot arm to move smoothly through a series of points in space, creating, in effect, continuous motion.

**MECHANICAL
TRANSFER DEVICES**

These non-servo devices are often called "pick and place" robots. They do not use feedback to go from one point to another. Only a limited number of end points can be specified. Each end point can be adjusted with great precision, but how the robot hand moves between points varies with friction, temperature, load and so on. Multiple points *are* possible, but the number is limited by the methods used to create robot "memory".

Pick and place robots are inexpensive, unsophisticated, but often very fast systems because they possess low mass. This accounts in large part for Japan's volume use of them. Pick and place robots often need more frequent "touch-up" adjustments than servo robots, due to mechanical wear as a result of system operation. This "touch-up", performed by maintenance personnel, adjusts the robot's delivery position or end point to the optimum location.

LIMITED SEQUENCE
ROBOTS

Use pick and place mechanisms to deliver parts from one place to another. They represent the low cost end of programmable robots. Unlike fixed automation, they can be adapted to different work situations. Some robot purists do not consider these devices robots. They do, however, represent a low cost answer to a large number of automation problems. These systems may use peg board memories or limited solid-state devices. Each motion can be powered electrically, hydraulically, or pneumatically. Pick and place robots generally require less horsepower than the more sophisticated servo systems.

PROGRAM CONTROL
SYSTEMS

The simplest program control systems are:

    1. timed sequence devices

    2. recorders

Timed Sequence Devices

Timed sequence devices were used on the earliest robots, and are still found on some pick and place units today. Timed steps sequence a series of preset control signals which are fed into the servo drive system. These command signals are fed to each axis of the robot or the wrist in timed increments. Sequenced in unison or one at a time, these timed signals create the point to point motion previously described. Timed sequence systems provide a simple and inexpensive means of producing complex motions in space. The major advantages of this control system are high reliability and low cost. The disadvantage is a cumbersome robot "teach" or program mode. However, for programs that are not changed frequently, the timed sequence control is an excellent cost effective solution to sequencing or programming a robot.

Recorders

Recorders were used on some of the earliest robot systems. The outstanding advantage of tape or disc recording systems is the teach mode. Continuous path signals are generated as the robot arm is moved through space. Playback merely repeats the space-time motions. A major disadvantage of this system is *lack of reliability.* Dirt on the record heads, tape wear, speed fluctuations of the recorder, and sensitivity to oil, mist, humidity, tend to limit this type of system to clean environmental conditions. For the most part, these systems are primarily of historic interest. If however, program changes are made frequently, recording robots might still be considered as a sequencing or programming system.

PROGRAMMABLE
CONTROLS

Programmable controls, a micro electronic growth area of today, can be used as a robot memory. Standard programmable units have been used in the process control and automation industries. Made primarily as standard control units for the process industry, the programmable control unit can be used as a low cost answer to robot sequencing.

COMPUTER SYSTEMS

The computer is the most sophisticated solution to robot sequencing problems. Most modern servo robots use computer systems. Capable of dealing with sensory feedback information, robots can begin to approach more human-like capabilities. By using computer systems, the robot manufacturer provides branching, multi program work selection options, end point error corrections, self-optimization programs, speed acceleration and mechanical wear correction options. The press of modern technology, through micro-miniaturization has forced computer memory cost and size to drop. At the present time, robot computers are housed outside the robot frame like the headless horseman. As the trend toward micro-miniaturization continues to evolve, and by improving environmental semiconductor stability, it won't be long before computer "brains" will be housed inside the robot. This will greatly assist the sensory feedback processes, it will minimize or eliminate rearrangement cost, and require less plant space.

SENSORY ROBOTS

Sensory robots are at the highest level of today's technology. This is one of the primary growth areas of robotics. Servo-computerized special feedback systems provide robots that use artificial sight, touch, and/or hearing. With one (or all) of these senses coupled into the work station robots can more easily duplicate human performance.

A great deal of research to date has been focused on artificial vision. But, sense of feel research is currently being accelerated. It is interesting to note that two-thirds of the human brain is taken up with tactile sensation. When you figure that the average surface area of a human being occupies 27 square feet of surface, and that there are an average of 300 sensors per square inch of surface, the staggering requirement of brain capacity becomes obvious. It is fortunate that a robot will not need tactile sensors over its entire surface. The data processing load would be staggering. By limiting tactile sense to the ability to do a specific job, the data processing load becomes manageable.

ARTIFICIAL INTELLIGENCE

This is the developing area of robotic technology. It makes use of pattern recognition and heuristic (trial and error) techniques. Computers programmed with "trial and error" programs can be used to simulate human decision making functions. "Mistakes" can be corrected. Parts that are not placed correctly on a conveyor belt, for example, can be correctly grasped, making the human robot interface (say in a loading situation) workable and desirable. Robots using these evolving developments will see, feel, hear and compare this sensory feedback information with patterns stored in memory. It will be said that these robots "understand". The field of artificial intelligence is rapidly expanding. I will cover this in greater depth in another section of the report.

ASSEMBLY ROBOTS

Another growth area in today's evolving robot field is the assembly robot. Computer controlled robots with sensory feedback can be used to assemble parts. Most robots are used to transfer parts, spot or arc weld, or to spray paint — tasks much easier than assembly. Assembly requires precision, or, as in humans, constantly correcting feedback. Assembly robots are useful for batch manufacturing applications where fixed automation is unaffordable. An assembly robot can be reprogrammed after each batch or can have a storage memory of previously progammed assembly tasks. Its cost is thereby amortized over a series of batch assembly tasks.

ADAPTIVE CONTROL

In baseball, adaptive control is what makes a great third baseman a great third baseman. In robotics, it is an option for some sophisticated systems which employ sensory feedback. Based on such feedback, whether visual or tactile, the robot corrects its path to conform to a shape or to a speed or to a position . . . as the third baseman does to make a catch. Adaptive control normally requires position, velocity, and acceleration feedback in addition to sensory feedback. Used in arc welding applications to correct welds as they are made, adaptive control is also utilized in the trimming or deburring of complex parts. Cincinnati Milacron, Automatix, Unimate, IBM, PUMA, Westinghouse and GE, to name a few manufacturers, each employ some form of adaptive control.

DIAGNOSTICS

"Physician heal thyself!" A current trend for the more sophisticated robot is a built-in computer program for malfunction diagnostic routines. This option permits company trained maintenance personnel to repair a robot, and also to minimize down time. Factory-trained maintenance personnel with sophisticated skills are generally necessary to effectively use the more complicated diagnostic routines. Some manufacturers offer telephone plug-ins direct from the factory to the manufacturer's computer which tell the operator what is wrong with his robot.

EDITING

After a robot is operating, an edit feature permits the operator to add, delete or replace specific motions (or program steps) and thereby to improve performance. The edit feature leaves the bulk of the program untouched. A useful feature is the ability to add sub-routines, so that the robot may perform secondary tasks while waiting between steps of its main routine.

ROBOT COORDINATE
SYSTEMS

1. Cartesian Coordinate
Robot

Cartesion Coordinate Robots (such as those manufactured by IBM and Cybotech), move in the classical three dimensional mode of in/out, up/down, forward/back. This robot motion system was among the last to evolve. The motions are shown in Figure 3.

2. Spherical Coordinate
Robot

This is the motion used by Unimate, the first robot system. It consists of the in and out axis (here called extend/retract) being swung up or down, with freedom to rotate. This coordinate

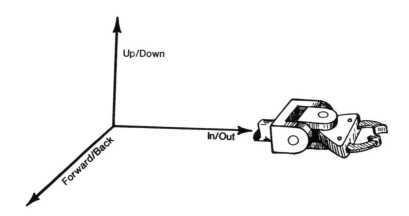

Figure 3. Cartesian Coordinate Robot

system is easily mechanized. It can be used in many applications, and provides a robot that can pick objects up from the floor (see Figure 4).

3. Cylindrical Coordinate Robot

Versetran, the second robot system evolved, employs a cylindrical motion. Here the x-y axis is rotated in space. Its primary advantage lies in the ability it provides the robot to keep an object moving in parallel motion with the floor (see Figure 5).

4. Anthropomorphic Robot

The anthropomorphic (literally, "human-like") design allows arm sweep, shoulder swivel, elbow and extension. This was one of the last systems to evolve. Milacrom and ASEA use these

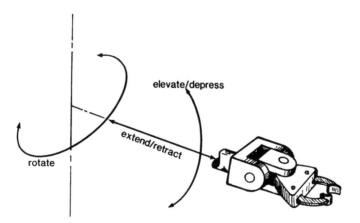

Figure 4. Spherical Coordinate Robot

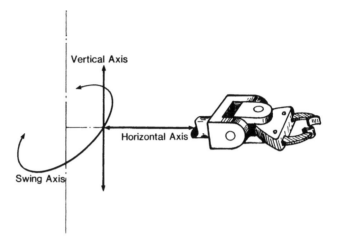

Figure 5. Cylindrical Coordinate Robot

movements. Advantages to anthropomorphism include an ability for the robot to go past its central axis without rotating around its mounting. Under some conditions reaching past the central axis can provide space-motion optimization (imagine reaching directly behind yourself for a tool; without turning around to get it!) (See Figure 6.)

GRIPPERS AND WRISTS (END EFFECTORS)

Grippers are a robot's hands. They can be built with many options. Simple open-close clamps are frequently used, or simple clamps utilizing snubbers to soften the clamping stress. Snubbers can use damping valves, rubber bumpers, or hydraulic or pneumatic deceleration devices. Other options include simple clamps with self aligning floating pads. These pads, with

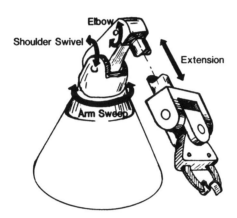

Figure 6. Anthropomorphic Robot

a built-in compliance factor, (for the pads float), makes them the best option for certain applications. It should be clear to the reader that grippers, or end effectors, are an innovative area. They often require special application design considerations. Other types of grippers include: simple clamps with special contours to fit specialty parts; sophisticated servo clamps with sensory feedback; suction cups for handling glass; chuck-like grippers that drill or rotate; magnetic grippers for special metals or scrap handling; shovel-like grippers for raw materials; and adhesive grippers.

Grippers often require special alignment for pick-up and delivery. This alignment is performed by the robot's wrist. Wrist and gripper combinations can be simple on-off devices or full servo mechanisms. Wrists are defined by roll, pitch and yaw as related to the robot's arm. For example, a six degree of freedom robot could have a spherical or cylinderical or XYZ coordinate system with a three degree of freedom wrist. Wrist options include:

Bang-Bang or settable two position. Wrists of one, two or three degrees of freedom are so named because of their here-or-there motion. Simplicity and their lightness of weight is their advantage—and a great one, when the power consumption costs of moving heavy servo mechanisms is taken into account. Their disadvantage is part wear and maintenance. Such wrists require frequent maintenance.

Servo Wrist. This can be either point to point or continuous; either way, it works on a feedback system. Options include:

- *One Servo Wrist Motion and Two Two-Position Wrist Motions*

- *Two Servo Wrist Motions and One Two Position Wrist Motion*

- *Finally, All Three Wrist Motions Are Servo Driven*

ROBOT WRIST MOTIONS

Robot wrist motions are akin to the motions of a battleship or an airplane, and are discussed in like terms of pitch, roll and yaw. They can be combined with gripper motions in any number of ways. Most grippers are of the bang-bang variety; yet grippers, since they come into contact with the work, are the one area of robots most often customized. Universalization, although badly needed, has yet to take place; each manufacturer works to his own standards. (See Figure 7.)

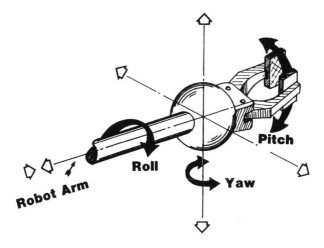

Figure 7. Robot Wrist Motions

INDUSTRIAL ROBOT SENSORS

Tacile sensors are important for assembly and for quality. Sensors useful in these areas include *pressure sensors,* in single or multiple arrays used for part orientation and gripping force control; *piezoelectric film* or *crystals* for continuous pressure; *conductive rubber* for tactile feedback (these sensors demonstrate high hystersis); cat-like *whisker sensors* to determine part presence or absence; *air jet proximity switches; ultra sonic* range finders; and *multiple on-off switches* spread over a surface.

Other sensors include *microphones* (sound sensors) and *optical sensors,* both single and multiple. The latter are chiefly of three kinds: either *interrupted beam* (as in automatic elevators), *Infra-Red* (IR), or *television scanning.*

Safety sensors are useful to prevent collision of the robot arm with objects, people, or other robots. These can be sophisticated ultra-sonic optical sensors, or simple surface switches. Optical curtain or beam triggers can zone off an area that is dangerous to personnel, and stop the robot in mid-cycle if a person walks into the "danger" area.

Displacement, velocity and *acceleration* sensors have been with us for years. They can be incorporated into a robot feedback system to provide stable position information.

HOME POSITION

Not every robot has a "home position", but most do. Simply defined, the home position is the place where the robot "goes" under special conditions. These conditions include: an out of the way reference start or end position; a special place where the robot goes under specific malfunction conditions. For example, if the equipment that the robot is servicing malfunctions, the robot arm can go to its designated home position. In some installations, getting to the home position can be tricky. If the robot has not malfunctioned, it can get there by way of a prescribed path. If the failure is in the robot, automatic shutdowns and manual manipulation are required to get the robot home. Non servo robots often use a home or reference position after each cycle to synchronize all axes that could otherwise accumulate errors. By using a home reference, inaccuracies in a cycle will not continue to build up.

MANUAL MODE

In manual mode, the robot moves in response to an operator's commands, or in a step by step operation as triggered by the operator. Such operation is analogous to power steering, where the servo system is continually nulling an error signal.

Manual mode can be used for maintenance checks, for trouble shooting, or as a means of disengaging the robot from other equipment under failure conditions. Manual mode is used in some robots as part of the "teach" or "program" system.

**TEACHING
MODE**

Some robots use a teaching mode that works like a power steering unit. A device clamped on the robot arm, often called a pendant, is used to steer it through a routine that is recorded in a computer memory on tape or disc. In some units, the operator commands the robot through a control console with push buttons or through a computer keyboard. Some anthropomorphic robots have a teach mode that permits conversion to rectangular or cylindrical coordinate systems. In these units, the operator selects the coordinate system prior to "teaching". This feature spans the best of both worlds. It permits the operator to use physical motions for programming which can be "cor-

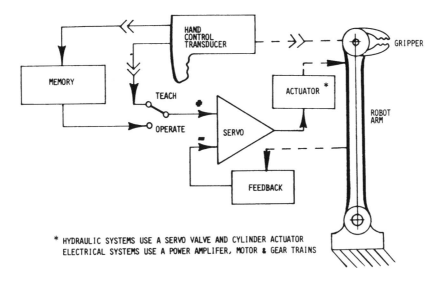

Figure 8. Teach Mode

rected" to the precision of a computer keyboard position command. (See Figure 8.)

**TRACKING A
MOVING LINE**

This option gives a robot the ability to work on parts or pick them up as they move along a conveyor line. Cincinnati Milacron for example builds a spot welding robot that can place hundreds of welds on a car body as it moves along a conveyor. As the line speeds up or slows down, the robot changes speed to place the spot welds properly.

OFF LINE
PROGRAMMING

Computer memory robots can often be equipped with off line programming. This is generally done in a remote programming center. The off line programming option permits new programs to be written for new production activities or for modifications of existing programs without interrupting ongoing production. Old programs can be kept in a file for reuse.

PROGRAM
INTERACTIONS

This option permits robots to deviate from a program in process and to branch into sub-routines. It also permits a robot to work in synchronization with other robots. Typical interactions include:

Periodic Self Maintenance. In arc welding, the robot can sense a worn tip during the cycle and it can change the welding tip as frequently as is needed.

Tool Change or Sharpening. The robot can sense a worn tool bit during operation and can branch to a tool sharpening or tool change routine.

Two Robots welding different ends of a part. By synchronizing each robot's motions it is possible to minimize heat distortion, with program interactions between the two robots.

Picking up parts that are not preoriented. With vision systems the robot can align itself with the part.

Performing miscellaneous sub-routine tasks while waiting for machinery or for other operations to reach completion. Unimation employs a concept known as flexible automation: if six machines perform a task, and one breaks down, a sub-routine can be switched in which will "skip" the malfunction, thereby keeping some production flowing. This contrasts with fixed automation, where everything is up and running or everything is down. Flexible automation is at the heart of robotics.

Branching into a search mode. This option permits the robot to look for parts or to stack parts while the machines that the robot is serving (i.e., NC machines, or Die Casting machines) are completing their duty cycle.

SCALING

A word used by some (not all) robot manufacturers, scaling is a feature which permits robots to function at multiple predetermined settings. *Speed scaling,* for example, is a velocity feedback system through which the speed of the robot at various points in its trajectory can be both controlled and monitored. Maximum velocity can be set to prevent excessive wear or power consumption. *Acceleration scaling* prevents overstressing or breakage of parts being handled, or of the robot itself.

STRAIGHT LINE OPERATION

Through computer calculations, anthropomorphic robots can be made to operate along straight lines. Straight lines generated by means of this computer programming option can be used to guide point to point robots along a "smooth" trajectory, without slewing. Many of the more sophisticated computer robots offer straight line operation as an optional subroutine, or as a standard built-in programming feature. In some of the computer/microprocessor controlled robots it is possible to pick points in space and then specify a "controlled path" between these points.

Artificial Intelligence (AI)

Having previously introduced the topic of Artificial Intelligence (AI) on page 38. I will use this section of the report to discuss the subject in greater depth. AI is a computer based technology that impacts several fields. Robotics is one of the fields that makes use of very limited AI capabilities as of this writing. AI is however, one of the growth areas of expanding robotic technology.

To better understand the limited use in robotics, I will briefly outline Artificial Intelligence research and applications. With a better understanding of the field, we will be able to understand why robotics cannot capitalize on current AI technology, and why it will eventually be the dominant robotic technology of the future.

53

In 1981, at a Dartmouth gathering, AI celebrated its silver anniversary. During its twenty-eight year existence (as of this writing), AI has had several names. These include *"Heuristic Methods"*, *"Complex Information Processing"*, and *"Automata Studies"*. Artificial Intelligence computer programs require:

1. A high speed reference memory system that contains data about the specific problem or a desired optimization system.

2. A Heuristic trial and error or self optimizing computer program. The program must:

   a. Accept input data

   b. Compare this data with the reference memory

   c. Use the determined procedures to solve problems or improve performance over known data

   d. Compare outputs to determine if the situation problem or performance was improved

   e. Store the comparison in the reference memory

   f. Take action based on the comparison.

3. Up date and streamline the reference memory to improve performance of the original program for the next problem or procedure.

From the above it is fairly obvious that "solving problems in real time" is difficult for the present state of the art computers. Problems that can be solved "off line" are easier to do than problems that need answers *now*. To speed up the process, special computer languages that are specific to AI heuristic usage are being developed. Computer memories and micro-processor chips with expanded capacity are also in development. The micro-electronic industry is currently manufacturing 256 kilo bit chips with random access memory (RAM) capabilities. Some manufacturers are considering layering the chips on one another to produce a 256 kilo bit $\times$ 256 kilo bit cube. Such a micro-electronic device would be capable of processing 65.5 billion* pieces of information.

Production of 256 K RAMS (kilo bit random access memories) is scheduled for 1983 and layered cubes will be available by 1990. Thus, a micro-computer of the 1990s will be more powerful than the mainframe computers of today. As these computer developments enter the field, and as AI research progresses, significant changes in robot technology will become available. Robots that truly see, hear, and perform a process akin to *thinking* will probably be available by the turn of the century.

Other AI programs for "problem solving" applications are currently being studied worldwide. Robot-only application of AI to such problems as *visual and tactile sensation speech recognition, language processing, automatic programming* and *assembly systems* are currently underway. Indeed, both Computer Aided Design (CAD), Computer Aided Manufacture (CAM), and Integrated Computer Aided Manufacture (ICAM)

---

*In the American system a billion is one thousand million.

are becoming integral parts of modern production systems. CAM and ICAM allow Research and Development people to proceed directly from an idea through computer documentation to working models. All the projects mentioned, and more, are part of today's research. Some of the major AI robotic research is listed in Table V "Artificial Intelligence Research in Process". Both Robotic and Non Robotic research is displayed in Table V. Some of the Non Robotic research being done at Stanford includes: The Dentral project; The Molgen project; and the RX project.

The Dendral Project is an AI program that helps organic chemists develop organic structures with the aid of a computer CRT screen. By computer simulation, the image of an organic structure can both be developed and experimented on, saving the scientist some expense and valuable laboratory time. The Molgen Project is another computer program currently assisting molecular geneticists with protein manipulation; and still another, the RX Project assists in the cataloguing and treatment of chronic diseases. The program assists doctors in evaluating symptoms to help them properly diagnose rare diseases. These applications of Artificial Intelligence have nothing to do with robots, but techniques learned in their development will eventually spill into the robot field.

In addition to its uses in the treatment of human malfunctions, A.I. can be used to check robotic "diseases". A computer diagnostic project is currently in the works, through a collaboration between IBM and independent scientists, which would examine and diagnose malfunctioning computers. This Summary of Artificial Intelligence shows where the Robotic field and modern technology are headed.

TABLE V

ARTIFICIAL INTELLIGENCE RESEARCH IN PROCESS

#	RESEARCH	Carnegie-Mellon U. Pittsburgh Pa. 15213 (412) 578-3826	Chrysler Institute 1200 East McNichlos Highland Park Mich. 48203 (313) 252-7150	George Washington University 725 23rd St. N.W. Washington D.C. 20052 (202) 676-6083	Hughes Research Lab. 3011 Malibu Canyon Road Malibu, Ca. 90265	Jet Propulsion Lab. 4800 Oak Grove Dr. Pasadena Ca. 91109 (213) 354-6101
1	Robotics	■				■
2	Tactile	■				■
3	Tactile Recognition	■				■
4	Vision	■		■		■
5	3 D Vision	■		■	■	
6	Image Recognition	■				
7	Computer Graphics					
8	Speech Processing					
9	Mfg. Technology & Assy.	■	■			■
10	CAM - CAD	■	■			■
11	Factory of the Future	■	■			■
12	Inspection	■				
13	VLSI Design					
14	Computer Architecture					
15	Human Memory					
16	Problem Solving	■			■	
17	Real Time Pblm. Slvg.				■	
18	High Level Robt. Lang.					
19	Mobility (Legs)	■				
20	Mobility Other	■		■	■	
21	Medical A.I.					
22	Computer Diagnostics					
23	Space A.I. Technology					
24	Heuristic Programming	■				
25	General A.I. Research					
26	Micro-Surgery	■				
27	Man/Machine Interact.	■				
28	Mining					■

ROBOTIC (topics 1–20)

OTHER (topics 21–28)

57

TABLE V (CONTINUED)

#	RESEARCH	Rensselear Poly-technic Institute Mfg. & Productivity Center Troy, N.Y. 12181 (518) 270-6724	NASA Marshall Space Flight Center Alabama 35812 (205) 453-2653	U. of New Mexico Alburqerque New Mexico 87131 (505) 277-6298	MIT 545 Technology Sq. Cambridge Mass. 02139 (617) 253-6218	Mfg. Productivity Center 10 West 35 Street Chicago Ill. 60616 (312) 567-9800
1	Robotics	●	●	●	●	●
2	Tactile			●	●	
3	Tactile Recognition				●	
4	Vision				●	
5	3 D Vision				●	
6	Image Recognition				●	
7	Computer Graphics	●			●	
8	Speech Processing				●	
9	Mfg. Technology & Assy.					●
10	CAM - CAD	●				
11	Factory of the Future	●				
12	Inspection	●				
13	VLSI Design	●				
14	Computer Architecture				●	
15	Human Memory					
16	Problem Solving				●	
17	Real Time Pblm. Slvg.				●	
18	High Level Robt. Lang.				●	
19	Mobility (Legs)					
20	Mobility Other		●			
21	Medical A.I.					
22	Computer Diagnostics			●		
23	Space A.I. Technology		●			
24	Heuristic Programming			●	●	
25	General A.I. Research			●	●	
26	Micro-Surgery		●			
27	Man/Machine Interact.					
28	Mining					

ROBOTIC (rows 1–19)

OTHER (rows 20–28)

Research Center	1 Robotics	2 Tactile	3 Tactile Recognition	4 Vision	5 3 D Vision	6 Image Recognition	7 Computer Graphics	8 Speech Processing	9 Mfg. Technology & Assy.	10 CAM - CAD	11 Factory of the Future	12 Inspection	13 VLSI Design	14 Computer Architecture	15 Human Memory	16 Problem Solving	17 Real Time Pblm. Slvg.	18 High Level Robt. Lang.	19 Mobility (Legs)	20 Mobility Other	21 Medical A.I.	22 Computer Diagnostics	23 Space A.I. Technology	24 Heuristic Programming	25 General A.I. Research	26 Micro-Surgery	27 Man/Machine Interact.	28 Mining
Univ. of Alabama Huntsville Alabama 35899 (205) 895-6015	■	■							■	■		■																
Stanford University A.I. Laboratory Computer Science Dp. Stanford Ca. 94305 (415) 497-2791				■	■	■	■	■		■				■	■	■	■				■	■	■		■		■	
SRI International Robotics Dept. A.I. Center 333 Ravenswood Av. Menlo Park Ca. 94025 (415) 859-2575	■	■		■	■			■			■	■			■	■	■							■	■			
Space Studies Inst. P.O. Box 82 Princeton, N.J. 08540 (609) 921-0377	■																			■			■			■		
SME 1 SME Drive P.O. Box 950 Dearborne, Mich. 48128 (313) 271-1500									■	■																		

RESEARCH CENTER ⇧

RESEARCH ⇨

ROBOTIC   OTHER

59

TABLE V (CONTINUED)

Research Area	Yale University Computer Science Dp. P.O. Box 2158 Yale Station New Haven Conn. 06520 (203) 436-8160	Veterans Administ. 3801 Miranda Av. Mail Stop 153 Palo Alto, Ca. 94304 (415) 493-5000x5465	Univ. of Southern California University Park La. Ca. Computer Science Dp. (213) 743-5501	Univ. of Rhode I. Dept. Elect. Eng. Kingston, R.I. 02881 (401) 792-2514	Univ. of Central Florida College of Eng. P.O. Box 25000 Orlando Fla. 32816 (305) 275-2236
1 Robotics		■	■	■	■
2 Tactile		■		■	
3 Tactile Recognition				■	
4 Vision			■	■	■
5 3 D Vision			■	■	■
6 Image Recognition			■	■	
7 Computer Graphics		■	■	■	■
8 Speech Processing					
9 Mfg. Technology & Assy.					
10 CAM - CAD					
11 Factory of the Future					■
12 Inspection					
13 VLSI Design			■		
14 Computer Architecture					
15 Human Memory	■				
16 Problem Solving	■				
17 Real Time Pblm. Slvg.					
18 High Level Robt. Lang.		■			
19 Mobility (Legs)					■
20 Mobility Other		■			■
21 Medical A.I.					
22 Computer Diagnostics					■
23 Space A.I. Technology					
24 Heuristic Programming	■			■	
25 General A.I. Research	■			■	■
26 Micro-Surgery					
27 Man/Machine Interact.					
28 Mining					

ROBOTIC (1–20) OTHER (21–28)

60

RESEARCH CENTER \ RESEARCH	1 Robotics	2 Tactile	3 Tactile Recognition	4 Vision	5 3 D Vision	6 Image Recognition	7 Computer Graphics	8 Speech Processing	9 Mfg. Technology & Assy.	10 CAM - CAD	11 Factory of the Future	12 Inspection	13 VLSI Design	14 Computer Architecture	15 Human Memory	16 Problem Solving	17 Real Time Pblm. Slvg.	18 High Level Robt. Lang.	19 Mobility (Legs)	20 Mobility Other	21 Medical A.I.	22 Computer Diagnostics	23 Space A.I. Technology	24 Heuristic Programming	25 General A.I. Research	26 Micro-Surgery	27 Man/Machine Interact.	28 Mining
Univ. of Cincinnati Dp Mech & Ind Eng Mail Loc 72 Cincinnati OH 45221 (513) 475-2738	■	■		■	■	■	■	■	■		■	■	■							■		■					■	
Naval Research Lab. Code 2610 Washington, D.C. 20375 (202) 767-3985	■	■		■	■	■										■	■			■								

ROBOTIC   OTHER

61

Robotics and Productivity

The following table examines the relationship between the percentage of GNP invested in capital (including robotics) and the consequent percentage of increase in productivity. Over the seventeen year study period, illustrated in Table VI, every nation examined has increased its total output capacity as a consequence of increased capital investment. While the United States remains the most productive nation in the world, it is important to note that *proportionally,* our increase over the 17-year period has been the smallest of any country studies — even smaller than the United Kingdom's. Note that the largest productivity increase — *one-hundred-seventy-percent* — occurred

63

TABLE VI

RELATIONSHIP OF PRODUCTIVITY AND INVESTMENT

1960 TO 1977 CUMULATIVE DATA*

COUNTRY	%GNP INVESTED IN CAPITAL	% PRODUCTIVITY INCREASE
CANADA	23%	28%
FRANCE	25%	65%
W. GERMANY	24%	78%
JAPAN	40%	170%
NORWAY	31%	40%
SWEDEN	22%	50%
SWITZERLAND	30%	61%
U.K.	20%	30%
U.S.	16%	18%

*MACHINE TOOL MANUFACTURERS DATA

in Japan, where the percentage of GNP invested in capital was the highest. Japan, a country of scarce natural resources and equally scarce manpower, can attribute a large share of its great gains in the world market directly to the intelligent use of capital investment; and a large portion of this investment has been in robotics.

Figure 9 gives us another perspective on the relationship between productivity and investment. In this graph where productivity increase is measured against total investment,

Japan's angle of increase provides graphic evidence that the more percentage of GNP a country re-invests in itself, the greater will be its output. Again the United States comes up with the poorest showing of all nations. The negative output of some of our most important industries, notably the steel and auto industries, accounts in part for this poor showing. Chiefly, though, we have been held back by our reluctance to re-invest; to modernize. It is ironic that scientific factory developments pioneered by the United States have been copied and perfected

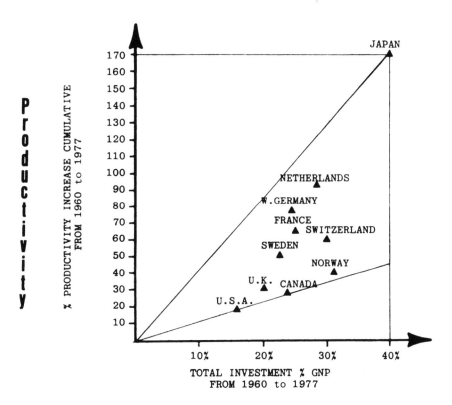

Figure 9. Relationship of Productivity/Investment from 1960 to 1977.

by the Japanese, to their advantage, while our own investors have dragged their feet, to our disadvantage. Inflation and the desire for fast returns have caused this problem to avalanche.

FACTORS AFFECTING PRODUCTIVITY

Table VII demonstrates that more than capital impacts on productivity. Over a third of the benefit of increased productivity is the result of technological development. Robotics is a combination of both technology and capital. It has the potential to impact productivity by 63.4%. Technology includes design change and modernization. Computers and microprocessors dramatize the impact of technology on output productivity.

TABLE VII

FACTORS AFFECTING PRODUCTIVITY

TECHNOLOGY	38%
CAPITAL	25.4%
LABOR QUALITY	14.2%
ECONOMICS OF SCALE	12.7%
RESOURCE ALLOCATION	9.5%

ROBOTICS AND AUTOMATION ARE CONTAINED IN TECHNOLOGY AND CAPITAL. JAPAN'S EXPENDITURES IN THESE AREAS HAVE CONTRIBUTED TO THEIR PRODUCTIVITY INCREASES.

Relay logic and for that matter circuit design has been virtually obsoleted by programmable controls, microprocessors and computers.

But other factors are important. Resource allocation accounts for just under ten percent of a country's productive potential. Considerations of resource allocation include the following considerations: Is power readily available? Is it inexpensive? Must raw materials be imported? While the United States enjoys a considerable advantage over most of its competitors, in this regard, we should note that an utter lack of natural resources has scarcely affected the Japanese.

Two related factors affecting productivity are quality of labor, and economics of scale. A learning curve enters into both. The more cars one makes, the faster a cost curve drops for a given technology and a given capital. Since GM makes twice as many cars as Ford, and Ford twice as many cars as Chrysler, GM has the best position against revenue loss. Although Chrysler's Engineers were among the very best in the automotive field the economics of scale and Japanese production systems helped cripple them.

Just as greater production means lower costs, so greater production can also mean *better products* — for the quality of labor, the human factor, is affected by a learning curve. If labor understanding grows, the quality of products also grows.

We can summarize these productivity factors thusly:

Improved technology, accumulated over many years, yields a 38% improvement in productivity. For a fixed tech-

nology, however, only investment of capital in modern machinery, plants, and automation coupled to the quality of human
labor, the economics of scale and source allocation leads to a
62% improvement in productivity. This is the area where Japan
has excelled; perfection of process and dedication to the work.

This brief glance at the human factor brings up a ques-
tion most often examined in relation to robotics, and that is
its potential impact on the American work force. Some people would have us believe that a cybernetic society is a worker's
purgatory, where a few wealthy citizens and their robot ser-
vants exist in luxury, while the masses starve or degenerate.
This argument has been used against technology since cave man
made an axe.

By historically reviewing the Bureau of Labor Statistics,
it is evident that fewer and fewer people are involved in the
force that produces. Our present Gross National Product is
produced by a work force in manufacturing, construction and
mining, agriculture, forestry and fisheries; some 33.8% of
America's population. The remaining 66.2% of people employed—and their number is growing—produce no tangible
wealth. These include public administrators, real estate and in-
surance agents, financial executives, business people, doctors,
nurses, and medical aides, transportation and welfare workers,
wholesale and retail workers. Ironically, it is in this indirect
segment of the economy that we find the "better life" of America's wage earners. At the end of World War II, the Blue Collar sector was a much larger percentage of the working population. Since that time, due to other forms of automation, the
Blue Collar sector has continued to decrease. (See Table VIII.)

TABLE VIII

1978 DISTRIBUTION OF EMPLOYMENT
(DATA FROM BLS)

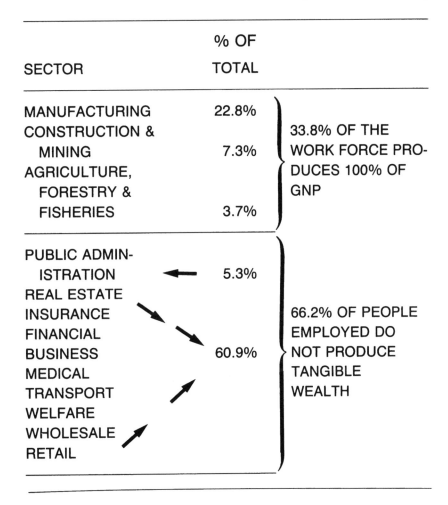

SECTOR	% OF TOTAL	
MANUFACTURING	22.8%	
CONSTRUCTION & MINING	7.3%	33.8% OF THE WORK FORCE PRO-DUCES 100% OF GNP
AGRICULTURE, FORESTRY & FISHERIES	3.7%	
PUBLIC ADMIN-ISTRATION	5.3%	
REAL ESTATE INSURANCE FINANCIAL BUSINESS MEDICAL TRANSPORT WELFARE WHOLESALE RETAIL	60.9%	66.2% OF PEOPLE EMPLOYED DO NOT PRODUCE TANGIBLE WEALTH

Thus, figures such as these suggest an equitable future can exist for our frightened workers. Robotics is the key to improved productivity, a key that will affect one-third of all employed Americans. A cybernetic society offers visions, not of breadlines, but of re-trained workers. Historically, as our productivity improves, our wealth grows, and with it our need for administrators and service sector personnel. We are presently living through the last phases of the Industrial Revolution, a revolution that will create artificially intelligent machines. This is something that Karl Marx could not visualize when Das Kapital was written. He could not realize that the ultimate end of the Industrial Revolution would be a society without workers in the classical sense. Socialism and the "Dictatorship of the prolitariat" seem to be a doomed society. It seems evident today that capitalism is the long term survivor, since trade and capital leverage are not concerned with how products are produced. They are concerned with trade and profitability.

In any event, a cybernetic society could mean more jobs for *people*. Some of those who now spend their days at labor drudgery could be re-trained. The balance will eventually flow into the service sector. Greater output production could lead to new and different work opportunities. Historically, this trend has been proven. The possibilities are unlimited. One of the world's first cybernetic factories, *Sara Lee,* automatically bakes cakes. This computer operated factory still uses people to fill the hoppers, to troubleshoot, and to keep the plant operating. Cybernetics means better jobs in the long term. During the transition, labor and management need to cooperate.

If we do not automate, if we continue with business as usual, we will not be able to compete on world markets. It is ironic that we, the inventors of robotics, find ourselves in this situation. If labor wins every battle against automation, they will ultimately lose the world battle and their companies will close. If management does not "help" labor over the transition, we will face riots and depression. The combative wound between labor and management must be healed. The alternative to cooperation is the decay of our society. We will become a "basket weaving" society with job loss and no replacement employment. Cooperation and intelligent application of robotics, can usher in the "second Industrial Revolution" gently.

ROBOT IMPACT ON PRODUCTIVITY

Robots improve productivity for the following reasons: They insure a lower cost per unit over manual operations for batch type work. Their production output is predictable, once set. Their quality is consistent: fewer rejects, less scrap generation, and saving in inspection are the result of Japan's experience with robot welders, and the experiences of some of America's automobile plants. Remember: In an inflationary environment, manufacturers must *improve* productivity just to stay level with inflation. To stay in one place is to fall behind.

The inflation rate over the past three years has averaged 11.6% in the U.S. At this rate of inflation, a company must double its sales every 6⅓ years just to stay level in constant dollars. Most highly capitalized companies have not attained this needed sales growth. They are, in actuality, shrinking in size. Many of these same companies have added employees during the past three years. Flat or shrinking sales in real dollars and increased employment are strong indicators of decreasing productivity.

Fortunately, many high technology companies have had growth rates that exceed inflation, and often with the same size work force. These companies continue to grow even in a recessionary environment.

The highly capitalized companies (i.e., steel and automotive) will only change the situation by reindustrializing their plants with automation, robotics, CAM (Computer Aided Manufacture) and ICAM (Integrated Computer Aided Manufacture). World wide competition will force this issue if these segments of our industry are to survive.

Economics of Robots

Most American companies have not met the inflation test of doubling sales in the last 6⅓ years. (To stay level in real terms.) Market segments have been shrinking, while employment remained stagnant or increased. This picture shows a national average toward decreasing productivity, which led to massive layoffs in the steel and automotive industries. If American industries are to survive or grow, we must reevaluate our use of capital and people.

Japanese strategists take a long time view of their customers, their competitors, and their company. In general, they will work on small competitive advantages. Their philosophy is to start small, but attain world domination. They are not interested in short time profitability. The Japanese use all of their resources starting with people, evolving into capital in the form of factories and machines, and finally into an examination of cash flow. Kenichi Ohmae, in his book "The Mind of the Strategist, The Art of Japanese Business", describes the CASIO companies philosophy of accelerating product life cycles to dominate a market. Obsolete your own products to beat the competition.

In dealing with highly competent world business competitors, our survival depends upon equally competent responses. We can no longer count on ten year depreciation cycles for machinery, and forty year depreciation cycles for factories or processes in *all* industries. Instead of having a single philosophy for business, our survival depends on multiple strategies aimed at markets, customers, competitors, and company objectives. The economics of robots are one of these strategies. Other considerations include manual labor and fixed automation.

WHERE ROBOTS PAY OFF TODAY

Each installation must be considered on its own merits. Depending on the degree of automation already in existence in the factory, and the number of cumulative products to be produced over the product's life cycle, it is possible to make product by product assessments.

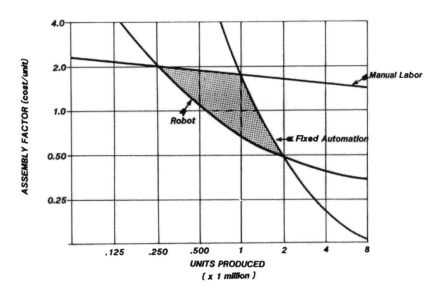

Figure 10. Where Robots Pay Off Today

In Figure 10, the three factors, people, robots, and fixed automation, are plotted on log-log paper for a specific installation. The manual labor learning curve is plotted for an existing facility. Robots, or flexible automation, begin to pay off at ¼ million units for this particular product line. (Other installations would be scaled differently.) Fixed automation in this installation pays off above two million units. Each manufacturing process and product can be examined for its break points. In some industries, at todays technological level, it would be foolish to use anything but people today. In many batch type operations, robots are the optimum choice. When product volume is very high, it is not possible to beat fixed automation.

FINANCIAL
ANALYSIS
OF A ROBOT

When in doubt, the easiest way to determine whether a robot is financially justifiable, is to evaluate the installation using Table IX. After reviewing the table and its use, we will perform a robot cost justification study for a specific installation.

In Table IX A, all of the cost factors, positive and negative, are examined. Naturally, before a cost verification can be made, a specific robot must be chosen. In subsequent sections, we will list Robot Manufacturers and their current cost ranges. Justification starts with item 1, the cost of the robot and its accessories. Accessories include conveyors, sensors, end effectors and grippers etc. Very often these costs can be greater than the cost of the robot.

Item 2 is maintenance cost and spare parts needed to ensure rapid repair when the robot needs maintenance. Each manufacturer recommends a list of spare parts.

Item 3 is the cost of robot down time relative to the production line. In some installations, this cost is negligible. In others, where a large number of machines are being maintained by a single robot, the cost per minute of down time can be very high.

Item 4, annual labor saving must consider all shifts. In some installations, a single shift will "pay off". In others, two or three are required. Annual labor saving can also include avoidance of accidents, reduction of workmen's compensation, and safety in general. In particularly hazardous environments, this factor can become the primary consideration. The company can not only save money, but the humane aspects of healthy workers, who remain undamaged by the work, goes beyond any balance sheet.

Item 5 considers the robot's ability to make a predictable constant quality product. Reduction of scrap is a balance sheet item.

Item 6, depreciation, is considered a cash inflow. Unfortunately, the fully paid back depreciated machine occurs at a different time than the investment. The depreciated money does not equal the invested money in real terms. By placing the item on a cash flow balance sheet, we can account for the time differences of money.

Item 7 considers the consumable tools used specifically for the robot installation. Special jigs and fixtures not needed by people must be accounted for.

Item 8 can be positive or negative. Robots use more energy than people do to do a job. The average person doing heavy work uses 1/20th horsepower. Many of the larger robots use 7½ to 15 horsepower to do an equivalent job. Smaller units of course use less power. The energy must be accounted for. It is important to note that the energy balance is not always negative since robots do not need light, or air conditioning, or special ventillation. People do. Conditions for a specific installation will determine if energy is positive or negative.

TABLE IX A.
FINANCIAL JUSTIFICATION OF A ROBOT

COST FACTORS		YEARS				
		1	2	3	4	5
1. ROBOT COST PLUS ACCESSORIES	(−)					
2. MAINTENANCE AND SPARES COST	(−)					
3. DOWNTIME COST	(−)					
4. ANNUAL LABOR SAVING	(+)					
5. QUALITY IMPROVEMENT AND SCRAP REDUCTION	(+)					
6. DEPRECIATION	(+)					

7. CONSUMABLE TOOLS   (−)

8. INCREASE OR DECREASE IN ENERGY   (±)

9. RESIDUAL NET ASSETS   (+)

10. CAPITAL DECREASES   (+)

11. INCREASED THROUGHPUT   (+)

12. TOTAL

13. DISCOUNT FACTORS@%

14. PRESENT VALUE

15. NET PRESENT VALUE

79

TABLE IX B.

FINANCIAL JUSTIFICATION OF A ROBOT*

COST FACTORS		YEARS				
		1	2	3	4	5
1. ROBOT COST PLUS ACCESSORIES	(−)	(216,200)	—	—	—	—
2. MAINTENANCE AND SPARES COST	(−)	(4,767)	(4,767)	(4,767)	(4,767)	(4,767)
3. DOWNTIME COST	(−)	(3,000)	(3,000)	(3,000)	(3,000)	(3,000)
4. ANNUAL LABOR SAVING	(+)	56,000	56,000	56,000	56,000	56,000
5. QUALITY IMPROVEMENT AND SCRAP REDUCTION	(+)	37,152	37,152	37,152	37,152	37,152
6. DEPRECIATION	(+)	12,880	12,880	12,880	12,880	12,880

7. CONSUMABLE TOOLS	(−)	—	—	—	—
8. INCREASE OR DECREASE IN ENERGY	(−)	(250)	(250)	(250)	(250)
9. RESIDUAL NET ASSETS	(+)	—	—	—	24,000
10. CAPITAL DECREASES	(+)	4,800	4,800	4,800	4,800
11. INCREASED THROUGHPUT	(+)	7,110	7,110	7,110	7,110
12. TOTAL		(106,275)	109,925	109,925	133,925
13. DISCOUNT FACTORS@%		1	.769	.592	.350
14. PRESENT VALUE		(106,275)	84,532	65,075	46,873
15. NET PRESENT VALUE $140,220					

*Negative values are shown in brackets.

Item 9, residual net assets, appear in the last year of the analysis. This sheet analyzes an installation over five years. If the depreciation of the installation is 10 years, the residual value at the end of five years is listed in the fifth year.

Item 10, capital decreases, may be large or small for a specific installation. If the installation uses less space and fewer machines than people or fixed automation, this can be a large ticket item. If, on the other hand, it uses the same space and equipment, this item is zero.

Item 11, increased throughput, is important. A robot feeding other equipment can speed up the process, making better use of existing capital equipment. I have seen automated welding installations that worked $4 \times$ faster than the original welders. Each installation must be measured on its own merits.

Item 12, is the total cost annualized for the installation. It is the sum of items 1 through 11.

Item 13, the discount factor is calculated from the cost of capital, or the hurdle rate a company needs to justify an investment. The discount factor is $(1/1 + R)^n$. "R" is the interest rate based on its cost of capital or its hurdle rate. In today's inflationary world, companies often set "R" at 30%. "N" is the number of years. Most investment calculations are based on discounting the reciprocal of compounding. This process brings all future costs or benefits into "today's" financial framework.

Item 14 multiplies item 12, the total cash flow, by item 13, the discount factor for each year. Item 14 is therefore the present value (or discounted cash) for each time period.

Item 15, is the cumulative summation of item 14. It represents the net present value of the five year calculation. If this number is positive, the company will make more than its "hurdle rate". I have seen robot installations where the "Internal Rate of Return" is over 60%.

ROBOT COST JUSTIFICATION STUDY

To illustrate the use of the table, we have filled out Table IX B for a robot cost justification study, we have chosen and assembly operation in an electronics manufacturing plant. A welded sub-assembly is used to house an electro-mechanical unit. The sub-assembly requires two seam welds. The data for this part is shown below.

- 206,400 sub-assemblies are produced each month.

- Three welders work to produce this quantity, two on the day shift and one on the night shift. During the day, two weld stations are used. At night, only one station is used.

- A fixture is used to hold the parts during the weld operation. Each welder produces, on the average, one part every nine seconds. (Their peak speed is one part every three seconds, but they cannot continue this rate all day.)

- The value added to the parts prior to entering this

station is $4.00. In an average month, the welders produce 5% scrap. (Occasionally they forget a weld or they produce a poor weld.)

- The welders' salaries including fringes and benefits are $28,000/year.

This sub-assembly operation was partially automated. A robot was used to do the seam welds. A man was used to load and unload a specially designed fixture. The fixture was designed to hold two parts in each of its four stations. Thus, the man loads eight parts in the fixture at a time. A rotary index table is used to hold two such fixtures. While the man is loading and unloading the fixture, the robot is welding. The fixture is designed for easy loading so that the man has a relatively low pressure job. The following data applies to the automated operation.

- The robot chosen was specifically designed for arc welding operations. Its cost, including an automatic welding control, was $117,200.

- Engineering design and development of the fixture, a rotary table, and conveyor system modifications, was $51,000.

- The fixtures, rotary table and conveyor modification costs, were $48,000.

Note: A good rule of thumb for estimating servo robot installation costs for accessories, design etc. is that these costs are approximately equal to the cost of the robot itself. Thus, for an installation that is being contemplated but that has not been designed, use a 2× cost multiplier. Conveyor costs predominate in pick and place installations, and for this simple type of robot the multiplier might be 3×.

- In this installation, scrap was reduced from 5% to below 0.5% with the robot welder. A common advantage of robot installations is constant quality.

- Depreciation is a source of cash only because it reduces ones' taxable income. For our calculation we will use 46% of the depreciated value (the current corporate income tax rate). To make the calculation simple, we will use straight line depreciations, but other methods which are in keeping with corporate procedures can also be used.

- We have chosen a hurdle rate of 30% for this calculation. Other rates which are in keeping with corporate policy can be used. Refer to the present value factor table attached. The table shows the present value of money at 30% interest. Thus, one year from today at 30% a dollar is worth .769 dollars, etc.

We are ready to calculate the net present value of this robot installation.

**Item 1. Robot Cost
Plus Accessories**

Robot cost $117,200

Engineering cost 51,000

Fixtures, Tables etc........................ <u>48,000</u>

Total Cost $216,200

Item 2. Maintenance and Spares

Maintenance personnel cost 38K per year (fringes and bene-fits included). We estimate twelve maintenance days per year. This cost is $1,767/year. Spare parts are estimated at $3,000/year.

Total Maintenance and Spares

Manpower $1,767

Spares .... <u>3,000</u>

Total Cost $4,767/year

An alternate approach to maintenance is through a service contract. This is a decision that can be made at the time of purchase.

Item 3. Downtime Cost

Very often installations are more reliable when robots do re-petitive work. Production is more predictable. It is possible that a robot production line will run more frequently with less down time than a production line run with people. For purposes of this calculation, we have assumed a negative impact of $3,000/ year, but modern robot installations could actually have a positive impact when compared to rest periods, holidays, sicknesses, and other human needs. .

Item 4. Annual Labor Saving

The robot in this installation saved the work of two men. Instead of three welders working on two shifts, one man worked on a single shift. He loaded and unloaded the fixtures on the rotary table. He was also trained to program the robot. Thus, the man/robot combination saved the company the annual labor cost of two men.

Annual salary for welders plus benefits and fringes for this company was $28,000/year. Thus the annual labor savings is $56,000.

Item 5. Quality Improvement and Scrap Reduction

The scrap rate with human operators was historically 5%. The value added for the parts when they reach this station is $4.00. With a robot welder, scrap is less than 0.5%. We will assume that scrap saving is 4.5%.

Monthly production rated 206,400

Quantity of parts scrapped* — 4.5% — 9,288

Value of parts salvaged by robot at $4.00 — $37,152

Item 6. Depreciation

For simplicity, we will use straight line depreciation, five years for the robot and ten years for the conveyor, fixtures etc.

Equipment cost robot	$117,200	5 year dep.	23,440
Fixtures, tables etc.	48,000	10 year dep.	4,800
Total Equipment	165,200		$28,240
Annual depreciation	28,240		
Income tax at 46%	12,880		

Enter the income tax value in item 6 starting in the first year, since we assume one year of work in the first year.

---

*Some of the scrap can be reworked, but the labor added and the extra handling for these parts were of almost comparable value. We have therefore assumed all of the part value as scrap to simplify the calculation. If rework is considerably lower in cost than the part value, the differential cost would have to be used in the calculation.

Item 7. Consumable Tools

Both the workers and the robot installation use the same quantity of consumable tools. For purposes of our calculations we will consider this a balance. In some installations, the robot system might use more tools or less tools than a human operator. If that is the case in your facility, put the value in item 7.

Item 10. Capital Decreases

This installation reduced the space required in the factory to manufacture the welded assemblies. The elimination of one welding station and the space required to operate the equipment was estimated at $4,800/year.

Item 11. Increased Throughput

In some installations, where manual labor creates a bottleneck, a robot may be able to work at the speed of other equipment. This speed up will produce more products. The value of this increase can be placed in the table in item 11. In one example we calculated an increased throughput of $7,110 annually.

Item 12

Item 12 is a summation of all positive and negative cash flows.

Item 13

Item 13 is taken from the present value table (see Present Value Factor Table). In this item we assume that the first year is this year and that the second year is the first discount year. If the first year of the project is next year (in the future), then we list the values exactly as they are in the table.

Item 14

To obtain item 14, simply multiply item 12 by item 13. Please note that negative cash flows are shown in brackets.

Item 15

Item 15 is obtained by adding all of the present value figures of item 14 together. The net present value is $140,220. This installation is an excellent investment. It yields considerably more

<u>Present Value Factor</u>

$(1/1 + R)^n$

HURDLE RATE 30%

1—.769
2—.592
3—.455
4—.350
5—.269
6—.207

than 30%. The internal rate of return is so high that for this installation, the cost of the robot and accessories could double, and the project would still meet corporate financial objectives.

Not all robot installations are as profitable as the one in this example. We have done many calculations that favor manual or fixed automation over robots. The key to this system's profitability is in the combination of manpower and quality improvement. Unfortunately, many companies forget to include the value of improved quality. It is a definite cash inflow. This installation would even meet corporate financial goals if only one man's salary were eliminated. Remember, for your plant or facility, that one of the side effects of robotics is consistent quality.

This calculation can be used as a model for assessing robot installations in your plant. The same form can be used to calculate other options (i.e., improvements to manual systems or fixed automation). By performing the calculation for several different options for a single installation, an optimum automation system can be chosen.

Labor's Reaction to Robots

Robots to date have relieved people of hazardous, dirty, and monotonous jobs. The replacement of American workers by robots has had small impact on the labor movement. Thus, today's labor organizations are not currently fighting robot installations.

Future, fully automated factories may be another thing. In order to prevent a hardening position, management must take the responsibility of retraining capable factory workers with updated skills. The new technologies require:

- technicians

- programmers

•skilled maintenance workers

•trainers

Phasing workers into these and other occupations can depolarize the situation. The Japanese transfer workers to other jobs when improved automation and robotics are installed in a factory. Japanese workers do not face job losses so they eagerly help implement automation concepts.

New labor contracts in the auto industry are making job security concessions to the United Auto Workers. An equitable solution to this problem for labor and management can help reindustrialize the U.S.

How to Specify a Robot

A robot is a "piggy-back" of servo systems. The gripper is carried by a wrist, the wrist by an arm with multi-coordinate systems; the arm finally rests on a pedestal which carries everything. Thus, *the major work performed by a robot is carrying itself.* Moving the payload is often the smallest part of the horsepower consumed by a robot. For example, a human being uses 1/20 horsepower on an exerting job. A robot on the same job might use over seven horsepower. It is therefore necessary that a robot be specified properly. To do this we must specify the load accurately. If improperly specified, extra money, poor availability, and extra energy will result. Robot users can develop a checklist that considers:

• the load and speed

- the accuracy needed

- the interrelationship of other machines, people, and operating shifts

- the wrist articulations and arm motions that will optimize the installation

- the frequency with which production changes are made

Problems will develop if maintenance and reliability are not considered *before installation.* To do this the user must review:

- environmental factors (temperature, dust, sanitation etc.)

- safety

- maintenance training requirements based on the robot's working system:

  - hydraulic

  - electrical

  - pneumatic

To prevent dissatisfaction, engineering and manufacturing are required to work together. Together, they can solve the corporate production needs for robots, fixed automation, and work force training.

How the robot uses energy is another factor to be considered for an installation. Energy considerations depend on the robots operating system. *Energy drain might be constant or intermittent.* For example, fixed hydraulic pump systems with accumulators can supply peak loads for short periods. The accumulator discharges during high demand and "pumps-up" during low demand periods. Horsepower drain is relatively constant over time. On the other hand, variable hydraulic supplies with small accumulators can conserve energy. This equipment is, however, more expensive at initial installation. By proportioning itself downward (giving out less energy when not in use), variable hydraulic supplies can ultimately pay back the extra investment via savings in energy.

Regeneration of energy occurs during the cycle with either hydraulic or electrical systems when heavy parts are moved downward. In essence, braking energy is returned to the hydraulic supply, or to the electrical lines. The motion of the robot arm itself "pumps up" the system, but the system, like a self winding watch, can only handle so much. An excess of regenerative energy can cause problems in some systems. This energy must have a place to go.

From an energy standpoint, *electrical* systems are the most efficient. At low levels of work, little power is drained from the lines, while at high work-levels power proportional to exertion is drained. Once again, regeneration "pumps up" the lines. But, because the motor, consisting of a roller and stator, sits on the robot's arm, extra weight is carried around. Electrical robots, unlike hydraulic robots, cannot be remotely magnetized. An hydraulic system permits separation of pump and actuator. Electrical systems often have the heaviest dead weight to load of any robot, since the stator must be mounted on the arm. In well-designed systems, this weight deficit can be partially offset.

Pneumatic systems are the least efficient from an energy standpoint, but this is greatly offset by their low "dead weight". These small, usually pick-and-place systems can often consume less power than an electrical or hydraulic system, since their low moving mass makes up for their energy inefficiency.

In general, no one robot system—hydraulic, pneumatic, or electric—offers the user a clear-cut advantage over all others. Each robot installation should be evaluated for the given application. To accurately predict performance, the anticipated application must be studied to determine which robot system is best for that application.

Table X is a technical summary of servo system comparisons. It can be used to help you assess a specific application.

Maintenance considerations include:

Routine Upkeep and Preventative Maintenance. This can be done on a contract basis with some robot manufacturers. Schools for training your personnel in standard repair procedures are also available. Robot repair can also be purchased from independent contractors.

Fault Diagnosis. Simple, built-in diagnostics exist in some robots. Additionally, some robot suppliers build in as op-

TABLE X

SERVO SYSTEM COMPARISON*

FACTOR	ELECTRICAL	HYDRAULIC	PNEUMATIC
1. BASIC SYSTEM	SOLID STATE LOGIC, POWER AMPLIFIERS, DC OR AC MOTORS, GEAR BOXES, BALL NUTS, COOLERS	PUMP, SUMP, REGULATORS (PRESSURE, TEMP., FLOW), FILTERS, HEAT EXCHANGERS, SERVO VALVES, MOTORS ACTUATORS & ACCUMULATORS	COMPRESSOR, INTERSTAGE COOLERS, PRESSURE CONTROLS, FILTER, DRYERS, MUFFLERS, VALVES, ACTUATORS SNUBBERS
2. WORKING FLUID	ELECTRICITY	HIGH QUALITY OIL BASE WITH ADDI-	AIR, NITROGEN, COMBUSTION

TABLE X (CONTINUED)

FACTOR	ELECTRICAL	HYDRAULIC	PNEUMATIC
		TIVES, WATER BASED SOLUTIONS, SYNTHETIC LIQUIDS	PRODUCTS
3. EFFICIENCY	OVER 90% FOR LARGE SYSTEMS*	SELDOM OVER 60%*	SELDOM OVER 30%*
4. SUSCEPTABILITY TO CONTAMINATION	LOW: (ELECTRICAL LINE NOISE ± 10%) EASILY HANDLED RFI NOISE EASILY HANDLED BY SHIELDING AND FILTRATION	HIGH: FILTERS REQUIRED SPECIAL HANDLING DURING MAINTENANCE. CLEANING PROCEDURES IMPORTANT. SERVO VALVES EASILY DAMAGED	INTERMEDIATE: LESS TROUBLE THAN OIL. PARTICLES DROP OUT BEFORE GETTING TO VALVES. MOISTURE & CORROSION A PROBLEM

5. DELIVERED WEIGHT TO FORCE RATIO	POOR-MOTOR & GEARING MUST BE CARRIED BY EACH SUB SYSTEM. POOREST WEIGHT TO FORCE RATIO	EXCELLENT-HIGHEST FORCE TO WEIGHT RATIO	FAIR-LIGHT WEIGHT, BUT LOW PRESSURES PRODUCE INTERMEDIATE FORCE TO WEIGHT RATIO
6. SAFETY OF OPERATION	SAFEST SYSTEM ELECTRICAL SHOCK HAZZARD & GROUNDING MUST BE CONSIDERED	LEAKAGE OF FLAMMABLE FLUIDS & FIRE HAZARDS HIGH VELOCITY JETS OF FLUID CAN PIERCE SKIN, BLOOD POISONING/EYE INFLAMMATIONS POSSIBLE	FLYING DEBRIS FROM RUPTURES CAN BE VERY DANGEROUS. EXPLOSIONS POSSIBLE WHEN VOLATILE OILS ARE PRESENT. (NITROGEN NON-EXPLOS.)

TABLE X (CONTINUED)

FACTOR	ELECTRICAL	HYDRAULIC	PNEUMATIC
7. TEMPERATURE SENSITIVITY & HEAT REMOVAL	LOW TEMPERATURE SENSITIVITY IN THE OPERATING RANGE, BUT POOR HEAT DISSIPATION RELATIVE TO HYDRAULIC	HIGH TEMPERATURE SENSITIVITY DUE TO VISCOSITY CHANGES. DIFFERENTIAL THERMAL EXPANSION CAN CAUSE TRANSIENT MALFUNCTIONS. EXCELLENT HEAT REMOVAL @ REMOTE HEAT EXCHANGERS	LOW TEMPERATURE SENSITIVITY. DIFFERENTIAL THERMAL EXPANSION CAN CAUSE TRANSIENT DIFFICULTIES. HEAT REMOVAL IS NOT NORMALLY A PROBLEM SINCE SYSTEMS VENT TO ATMOSPHERE
8. OPERATION AROUND NULL (THE QUIESCENT OR INACTIVE	VERY LOW QUIESCENT POWER DRAIN (BELOW 5%)	INTERMEDIATE QUIESCENT POWER DRAIN WITH CLOSE CENTERED VALVES	HIGHER QUIESCENT POWER DRAIN AROUND NULL WITH CLOSED CENTER

		(BELOW 10%) HIGH POWER DRAIN WITH OPEN CENTERED VALVES (80% DRAIN)	VALVES (AROUND 20%) HIGH POWER DRAIN WITH OPEN CENTERED VALVES (UP TO 100%)
STATE) THESE FIGURES ARE FOR A BALANCED CONDITION.			
9. INPUT POWER SUPPLY	24 VOLTS TO 460 VOLTS	50–5000 PSI	5–500 PSI
10. LOAD VARIATION SUSCEPTABILITY	DRY FRICTION ON OUTPUT MOTORS CAN CAUSE SMALL TO INTERMEDIATE STEADY STATE ERRORS@NULL!	DRY FRICTION ON OUTPUTS CAN CAUSE SMALL STEADY STATE ERRORS AROUND NULL.	SERIOUS STEADY STATE ERRORS CAN OCCUR AROUND NULL POSITION DUE TO DRY FRICTION ON OUTPUT SHAFT.

TABLE X (CONTINUED)

FACTOR	ELECTRICAL	HYDRAULIC	PNEUMATIC
11. COMPONENT AVAILABILITY	AVAILABILITY VERY GOOD & IMPROVING	EXCELLENT AIRCRAFT AND INDUSTRIAL COMPONENT AVAILABILITY	POOR AVAILABILITY OF SERVO COMPONENTS
12. SYSTEM STIFFNESS*	REASONABLY STIFF— STIFFNESS DEPENDENT ON SPEED REDUCTION CONCEPTS**	THE STIFFEST, MOST RESPONSIVE SYSTEM FOR HEAVY LOADS	VERY SOFT SYSTEM (HYDRAULIC IS 400 × STIFFER)

* A well designed electrical system can approach the stiffness of an hydraulic system. How the reduction, motor and arms is designed is critical.

**This assumes a balanced system. With "dead weight" each system must supply power to compensate for the imbalance.

104

tional equipment telephone modems for diagnosis. Ultimately, large, complex installations require the diagnostic services of trained technicians.

Availability of Equipment and Needed Delivery Time. Spare parts for frequently encountered problems will greatly assist robot availability. (Availability is the time that the robot is available for work as opposed to down time when it is being serviced.) Designers, engineers, and manufacturing personnel should work with the robot manufacturer. Before installing any robot, discuss part availability, manufacturer's warranty and liability and so on. In some installations, robots should be capable of being backed out of the way so that human operators can "pinch-hit" during serious down-times.

Repair can be tricky. Contamination of hydraulic systems can be serious, if unsophisticated maintenance personnel are used. (A speck of dust can ruin a $700 servo valve.) Likewise, electronic systems require adept technicians: microcircuits, for example, can be ruined if improperly shielded. Happily, many robot manufacturers use a modular approach to maintenance. With proper diagnostics, finding, removing and replacing the "trouble spot" *can* be a matter of minutes.

Remember that robotics is a new technology for most companies. *If robots are to be succesfully installed, engineering must work with manufacturing to take the first step!* Improved productivity can be *planned*. A "patch job" will generally prove to be unsatisfactory. An analysis of a company's needs will include:

•the company's basic requirements.

•the specific jobs that are best assigned to robots.

•the degree of sophistication that manufacturing, engineering, maintenance, and upper management have.

•the part weight, accuracy, and speed needed for each specific installation.

•the effects of initial robot down-time the first installation will have on production.

•the specific environmental conditions.

Robotics is not an answer for everyone. Before budgeting and specifying a robot, consider fixed automation and/or skilled workers. A sound evaluation will prove which approach is most cost-effective for your company! Well thought out robot installations will pay back dividends in better quality and more consistent production and will minimize transient human resource dislocation.

Robot Manufacturers

In 1983, the number of robot manufacturers offering products in the U.S. exploded in the marketplace. Seventy companies offer two-hundred-and-five different robot designs. Although many of the robots look different, they fall into approximately thirty classes. These classes are based on:

- Payload: The robot's ability to carry or support a given weight or load.

- Repeatability: The robot's ability to return to the same spot from a given direction and speed.

- Speed: The robot's ability to deliver its payload at peak speed. Many manufacturers specify no load speed, but it stands to reason that as the load is increased, the robot slows down. It is loaded speed that matters during delivery, and unloaded speed is important on the return.

• Reach: The robot's ability to work in a space envelope. The larger the envelope, the sturdier the robot must be. The larger the envelope, the more difficulty there is in maintaining its repeatability. Large envelopes and tight repeatability tolerances require sophisticated servo systems. Thus, robot cost goes up with increasing size and decreasing tolerances.

• Robot type: The robot's coordinate system, its power drive system, and its ability to maintain orientation. In many installations, the robot selected is based on environment, safety, and available space. These requirements can often limit the robot to a specific type.

• Normal Price: An industry average price for single quantity purchases.

• Normal Use: The way industry uses robots of this class.

• Number of Manufacturers: This shows how many manufacturers offer the class of robots defined by payoad, repeatability, etc. Class is independent of the design appearance. Many robots have a different appearance, but their function and ability is the same.

The data in Table XI are an average of several different manufacturers products. Each column therefore, represents average data for the class. For example, when we list 8 to 30 inches for a 2.2 lb. cylindrical pneumatic robot, the figures represent what most manufacturers provide in that class. It is possible to find a manufacturer with a 32-inch reach within the class. The dollar figures are also averages. For example, one manufacturer of a one lb. pick and place electric robot will sell the unit for $2,300 while another will sell a similar robot for $2,500. To get the exact data, you will have to work with the robot manufacturer.

Our purpose in putting together Table XI is to show the range of available robots, their average price and performance characteristics. By matching your installation needs of payload, speed, reach, and position repeatability with the tabulated data, you will be able to determine the average cost and sophistication required for a robot installation in your plant.

To aid your search, we have provided a separate insert that tabulates robot suppliers in the U.S.A. If you are located in a country outside the U.S.A., we advise that you contact your local robot society for a list of available manufacturers.

MANUFACTURER'S PHOTOGRAPHS

The following photographs have been included in this book to show the range of representative robots and devices available in today's market.

Figure 11   Unimate Hydraulic Servo Robot, Light Trace

TABLE XI

REPRESENTATIVE ROBOT OFFERINGS IN 1984

Payload	Repeatability in Inches & Speed	Reach Min to Max	Robot Type General Description	Normal Price Options	Normal Use	Number of Manufacturers of the Class
1 Lb.	±.030/7 ips	6" to 15"	Anthropomorphic electric stepper motor, pick and place, non servo, 3 axis	$2,400 without control	small parts transfer	13
2.2 Lb.	±.001/40 ips	8" to 30" limited vert. to a few inches	Cylindrical pneumatic limited motion, pick and place non servo robot, 3 axis	$5,500 without control	small parts transfer & limited assembly	16

TABLE XI (CONTINUED)

Payload	Repeatability in Inches & Speed	Reach Min to Max	Robot Type General Description	Normal Price Options	Normal Use	Number of Manufacturers of the Class
5 Lb.	±.005/20 ips	10" to 25" 270° sweep	Spherical, pneumatic pick and place non servo robot, 5 axis	$10,000	loading unloading & parts transfer	2
2.2 Lb.	±.002/39 ips	7" to 25" 200° sweep	Anthropomorphic Electric servo robot, 4 axis sensory feedback	$30,000	precision small part assembly & transfer	5
5 Lb.	±.004/55 ips	8" to 16" 355° sweep & 300° vertical	Spherical, electric servo robot, 5 axis sensory feedback	$37,000	small part assembly & transfer	2

6 Lb.	±.040/20 ips	4" to 34" 320° sweep	Anthropomorphic electric servo robot 6 axis	$41,000	small parts assembly & transfer	3
11 Lb.	±.002/20 ips	6" to 25" 270° sweep & 30" vertical	Anthropomorphic electric servo robot 6 axis	$30,000	precision small parts assembly	4
14 Lb.	±.001/30 ips	51" by 12" by 10" arranged in banks	X,Y,Z. coordinate electric servo robot 12 axis up to 4 arms	$125,000	sequential small part component assembly	2
16 Lb.	±.008/40 ips	0 to 58" set within a frame	X,Y,Z, coordinate hydraulic servo robot multi arm, 6 axis	$100,000	sequential small part assembly	4
22 Lb.	±.006/40 ips	3" to 67" & 285° sweep 0 to 80" vertical	Anthropomorphic electric servo robot 6 axis	$65,000	assembly & transfer machine cell	13
35 Lb.	±.008/60 ips	50" to 120" & 80" by 80"	X,Y,Z, coordinate electric servo robot 5 axis	$135,000	Welding parts transfer & assy.	12

TABLE XI (CONTINUED)

Payload	Repeatability in Inches & Speed	Reach Min to Max	Robot Type General Description	Normal Price Options	Normal Use	Number of Manufacturers of the Class
50 Lb.	±.040/80 ips	20" to 78" & 360° sweep	Anthropomorphic pneumatic servo robot, 5 axis	$16,000	part transfer, machine loading	1
50 Lb.	±.050/30 ips	15" to 50" & 270° sweep	Cylindrical, hydraulic servo robot, 4 axis	$28,000	Parts transfer, & crude assembly	8
50 Lb.	±.050/30 ips	47" to 80" +30°, −28° in vert. & 208° sweep	Spherical, hydraulic servo robot, 5 axis	$30,000	parts transfer, & crude assembly	1

114

Payload	Repeatability	Motion	Type	Price	Application	No.
70 Lb.	±.010/25 ips	15" to 99" hor., 0 to 130" vert. & 270° sweep	Anthropomorphic electric servo robot 6 axis	$80,000	Part assy. & transfer machine loading	6
100 Lb.	±.008/15 ips	20" to 42" 10" vert. & 270°sweep	Spherical, Pick & place robot, 3 to 5 axis	$30,000	parts transfer	4
100 Lb.	±.004/15 ips	0 to 48" on X 36" on Y & 108 on Z	X,Y,Z, Coordinate electric servo robot 6 axis	$95,000	part assy., transfer, inspection	7
100 Lb.	±.050/50 ips	20" to 90" Hor., 0 to 154 vert., & 240° sweep	Anthropomorphic Hydraulic servo robot 7 axis	$80,000	heavy assy., paint spray, welding & part trans.	5
150 Lb.	±.010/30 ips	25" to 102" 0 to 154" vert. 240° sweep	Anthropomorphic hydraulic servo robot 6 axis	$90,000	part assy., inspection, machine cell	3

TABLE XI (CONTINUED)

Payload	Repeatability in Inches & Speed	Reach Min to Max	Robot Type General Description	Normal Price Options	Normal Use	Number of Manufacturers of the Class
175 Lb.	±.008/30 ips	15" to 60"X, 40"Y, & 40"Z	X,Y,Z, coordinate hydraulic servo robot 5 axis	$162,000	heavy assy., welding & part trans.	6
225 Lb.	±.050/35 ips	20" to 102" hor., 0 to 150" vert. & 270° sweep	Anthropomorphic hydraulic servo robot 6 axis	$85,000	Forging & heavy part transfer & crude assy.	2
250 Lb.	±.050/36 ips	15" to 50" & 270° sweep	Cylindrical, hydraulic servo robot, 7 axis	$72,000	heavy assy. Part trans.	2
300 Lb.	±.050/20 ips	38" to 71" hor., +30°	Spherical, hydraulic servo robot, 6 axis	$55,000	Part trans. spot weld &	2

Capacity	Accuracy/speed	Range	Type	Price	Application	Qty
		−28° vert. & 208° sweep			heavy assy.	
450 Lb.	±.080/20 ips	57" to 115" hor., +28° −23° vert. & 200° sweep	Spherical, hydraulic servo robot, 6 axis	$70,000	heavy duty part trans. forging	2
600 Lb.	±.050/20 ips	15" to 50" & 270° sweep	Cylindrical, hydraulic servo robot, 6 axis	$80,000	Heavy assy. parts trans.	1
2000 Lb.	±.080/15 ips	15" to 50" & 270° sweep	Cylindrical, hydraulic servo robot, 6 axis	$125,000	Heavy part trans	1
welding special	±.060/30 ips	10" to 45" hor., 80° sw. & 80° vert.	Spherical, hydraulic servo robot, 5 axis	$38,000	Arc weld	1
welding special	±.010/300 ips	50" to 120" hor., & 80" vert, by 80"Z	X,Y,Z, coordinate electric servo robot	$125,000	Arc weld fully equipped	15
special paint spray	±.032/36 ips	10" to 90" vertical & 93° sweep, & 71" horiz.	Anthropomorphic hydraulic servo robot. Robot fully equipped to paint.	$75,000	Paint spray	6

117

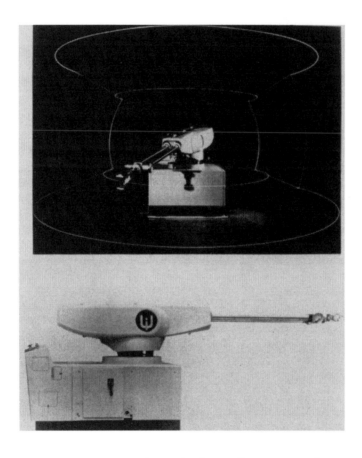

Figure 11. Unimate Hydraulic Servo Robot, Light Trace. This photograph shows the spherical geometry of this robot system.

Figure 12. Unimate Puma Assembly Robot. This robot is used for small parts transfer and assembly.

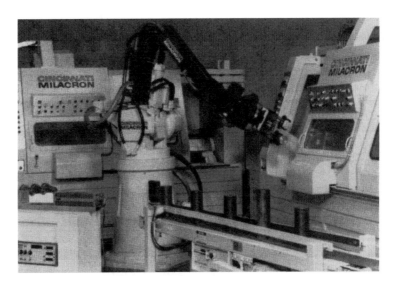

Figure 13. Cincinnati Milacron Hydraulic Robot N.C. Turning Machine Center. This robot is part of a manufacturing cell.

119

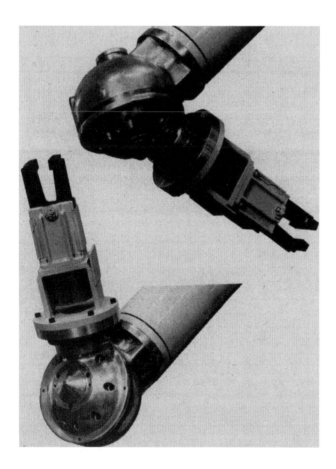

Figure 14. Cincinnati Milacron Three Roll Wrist. This photograph shows the flexibility of a three degree of freedom wrist.

Figure 15. Prab Model E and Movable Track Robot. This photograph shows a robot capable of moving along a track. It can perform work in two places and act as a conveyor system.

Figure 16. ASEA Robot Spot Welding Automobiles. This robot carries a 100 Lb. welder, while it places spot welds on an autobody with repeatable precision.

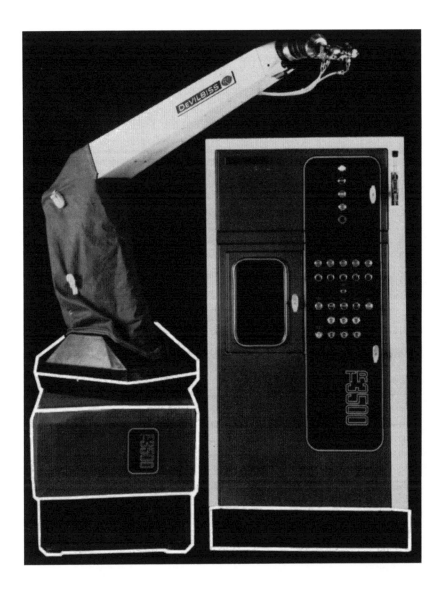

Figure 17. Devilbiss Paint Spray Robot. This photograph shows the control console and the paint spray robot.

Figure 18. IBM RS 1. This robot is a prototype of the advanced IBM 7565 Manufacturing System. It is controlled by a modified IBM Series/1 Computer.

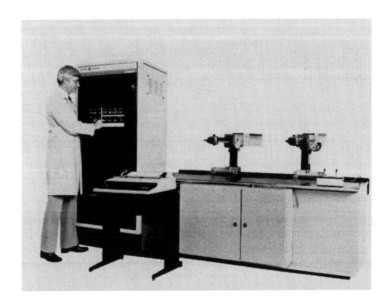

Figure 19. General Electric A12 Assembly Robot. The 812 is a unique, modular, Assembly Robot. It is available with one to four arms and up to twelve DC servo controlled axes.

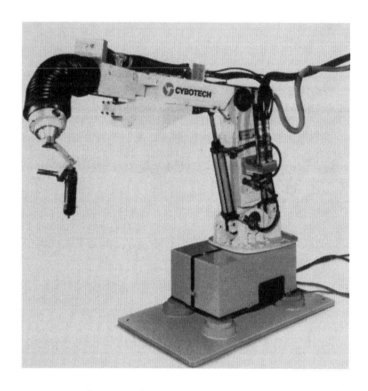

Figure 20. CYBOTECH Hydraulic Paint Spray Robot. This paint spray robot makes use of an unusually versatile flexible jointed front arm. The arm can curve in any direction to spray paint at difficult to reach positions.

Figure 21. Westinghouse Automation System Assembly Robot. This Unimation/Westinghouse Assembly Robot is part of a small parts assembly cell. The robot can have single or multiple arms.

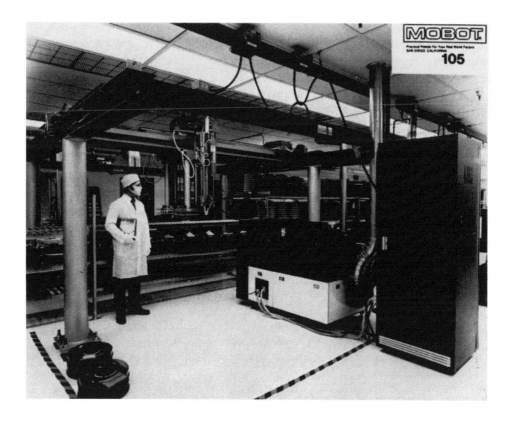

Figure 22. Mobot Computer Disc and Separator Ring Transferring
Robot.   This was especially configured for clean room service using
scavenged exhaust ports on the pneumatic cylinders and special low
vapor pressure lubricants.

Figure 23. SEIKO Pick and Place Robot. This pick and place non servo robot is used for part transfer or small parts assembly.

Robot Safety

This section of the book will discuss potential safety hazards encountered in the use of automated machinery and robots.

Fires are one of the potential hazards of high pressure hydraulics. High pressure hydraulic oil, when atomized through a leak the size of a pinhole, can create an explosive environment. It's a good idea to minimize the number of flexible hose lines used in hydraulic installations. If specifying a hydraulic robot, consider fire-safe procedures. Care and thought must be taken, however, as some obvious "answers" may not be answers at all. Fireproof fluids, for example, can often have undesirable properties in and or themselves. Often their lubrication qualities are poor, and additives in fireproof fluids can be poisonous. Each installation should consider the fire hazard in a flammability study. Procedures and fire prevention systems should be part of the planning. If properly handled, hydraulic robots can operate safely.

Electrical Systems can present different hazards. Grounding and shock prevention must be considered when evaluating an electrical system.

A recent newspaper headline started "Robot Kills Worker". In reading the article about a Japanese worker who violated all safety regulations, it became obvious that the headline was there to be an eye catcher. Robots are, after all, machines that at the present time do not contain intelligence. You don't read about lathes, milling machines, punch presses, numerically controlled machines killing workers. Yet, many workers die or are maimed while operating these machines. When it comes to robotics, there is an emotionally charged atmosphere. Are we still fighting the Frankenstein myth? I prefer Isaac Asamov's three laws of Robotics. One of the inherent built in non-removable commands is that a robot cannot harm people.

We, the robot designers and applications engineers, are after all in charge of the situation. We can program robots to perform set routines when safety violations occur. Fortunately, many robots today have built-in safety monitors. These monitors can include position timing safety feedback information; visual monitors; and proximity monitors, which cut the power to a robot when a human enters the operation zone.

When it comes to safety, we must not take anything for granted. The safety aspects of a robot installation can be planned. The following list of *what if*'s is a checklist for safety planning. What If:

- A person is in an unauthorized area

- The robot fails

- Power fails

- The robot is caught in a press

- Hydraulic fluid is leaking

- The robot drops or hurls a part

- The control fails

- The robot is in a hazardous environment where:

    - Fire

    - Nuclear leaks

    - Contaminated atmosphere

    - or Vacuum

  is a safety hazard

- Machinery around a robot fails

- The robot mountings fail

- Etc.

For every what if there must an answer. "Fail safe" devices, screens, guards, proximity detectors, alarms etc. can be used to answer each safety need. In some applications, hydraulic, electric, or pneumatic devices may not be applicable. Fortunately, when one device is ruled out others are feasible.

Many robots today have built-in safety monitors. These monitors can include position timing safety feedback information; visual monitors; and proximity monitors, which cut the power off when a human enters the operation zone.

Safety must not be taken for granted. In robot installations, as in any machine installation, it must be planned!

Robot Mounting Systems

Remember that if the robot mounting is unstable, the stated end point accuracy cannot be achieved. Most robots are floor mounted. Bolted to the floor, or to a mounting plate, a robot can be quickly disconnected to allow for human replacement during "down time". "A" frame floor mounts serve as a means of clearing floor space. They permit a robot to be mounted above the work area, thereby leaving a small "foot print" in congested areas. Specifically designed robots can be made on standoffs, pedestals, and "A" frames to clear floor areas.

Overhead, or side-hanging mounts, allow human replacement or servicing to be achieved easily. When a robot is malfunctioning, an operator simply pushes the robot out of the way. Such mounts, though, can have their accuracy compromised by building vibrations or by shifts in the mounting itself.

When true sensory robots are a working reality, mounting systems will be less critical. Visual and tactile feedback will help the robot correct for mounting shifts. For the present generation of robots, however, we must take special precautions when they are mounted on a moving system (i.e., the robot goes to the work instead of the work coming to the robot). To guarantee performance, we must make sure that the working positions are locked to a stable reference. This can be accomplished with special clamps and brakes.

Sophistication and Training

Hydraulic robots require skilled maintenance personnel. The robot can be converted into a pile of junk if hydraulic fluids are contaminated. Maintenance personnel must be trained to monitor filter plugs and warning lights. They should be schooled in the proper handling of servo valves. Such valves, though highly expensive, can, in the hands of unskilled technicians, act as "fuses" in a contaminated system. Oil change procedures, cleaning and refilling the sump, the use of thermal controls and the prevention of oil varnishing (due to overheating) all require skilled maintenance personnel.

Grounding procedures for servicing microelectronic equipment requires sophisticated technicians. In the hands of an unskilled technician printed circuit boards, computer subassemblies and power stages can be destroyed.

If maintenance is performed by "in-house" personnel, robotic down time can be reduced. It's a good idea to check warranties and liabilities, however, before assigning one of your own technicians to this task. If the manufacturer supplies maintenance, you should check for contract protection and emergency service. Actually, manufacturer supplied maintenance is a good way of getting started; but try to have one (or more) of your own people present so as to learn the ropes.

Note that trained maintenance people, while always crucial, are important for hydraulic and electric robots. Since hydraulic robots generally use microelectronic or computer feedback systems, service technicians need to understand both technologies. If all other factors are equal and either type of robot will suit your application, the single technology maintenance requirement of electric robots can be a deciding factor.

At the time you purchase a robot, you should have already checked maintenance availability. Explore the size and availability of each robot manufacturer's service personnel, reputation, recommended spare parts list, and service and installation contract. (Make sure, by the way, that there *is* a recommended spare parts list!) Talk to other robot users before you buy. The importance of proper maintenance *to the manufacturers themselves* is illustrated by the history of IBM and UNIVAC. Some years back, these two computer manufacturers were running neck and neck in the market. UNIVAC actually had a technologically more sophisticated computer. IBM concentrated on the development of a large service organization.

When the UNIVAC vacuum tube computer went down, it took a long time to get service personnel for repair. IBM, on the other hand, repaired their units in a short time. Today, IBM has far surpassed UNIVAC in computer sales. This story underscores the importance of service and maintenance to the reputation and growth of a company.

Some manufacturers run service maintenance training seminars. If a manufacturer whose product interests your firm offers such a seminar, by all means have one of your employees attend it. Your employee should examine the manufacturer's diagnostic procedures and aids. Does the robot manufacturer offer diagnostic plug-in modules? What trouble-shooting procedures are used? Are there special installation requirements, interlock requirements? What tools and equipment are required for routine maintenance? How frequent are maintenance checks scheduled? (Daily? Every hundred hours? Etc.) What are the robot's lubrication requirements, and what is its optimum lubrication schedule?

A final and crucial aspect of maintenance considerations is this rule, to be followed whenever possible: *Robots should be installed so that they can easily be serviced*. Preventive maintenance programs can minimize costly breakdowns; but installations that are poorly thought out can lead to assembly line shutdown and user dissatisfaction.

TRAINING

It should be obvious by now that training all levels of personnel is a must if robotic technology is to be successful. A trained organization will easily cope with robot problems. Untrained personnel can cause failures.

TRADE OFF STUDIES

If you plan to install a robot in your factory it's a good idea to conduct a trade off study. The study should consider: "payload" weight; speed; end point accuracy and repeatability; and system elasticity. Comparison of different robots for the same application, with fixed automation or upgraded workers can clarify the direction you should take. Overspecification of any of the installations real needs can lead to high cost and poor performance. Engineering and manufacturing can develop a trade-off study before a robot or its alternatives are purchased. It's a good idea not to simply replace people with robots. Considerations can include *product redesign for a robot*. If your product can be modified slightly or redesigned for ease of handling, a simple robot might be able to handle the job.

NEVER LET GO

One of the principles of automation is "Never let go of a part until you have completed *all* operations". In too many factories, machines drop parts into bins and on to conveyors. Then a search and align operation is necessary to further process the part. Proper alignment may require vibratory feeders, and mechanisms to properly orient the part, or the use of modern vision systems. The robot equipped with a vision system can determine part orientation, and align its gripper to properly pick up a part. In many instances all of this is unnecessary. Simply transferring the part from one machine into an aligned receiver or into another machine, can speed the process and save money.

.

Conclusion

It is fairly obvious today that the "Factory of the Future" will use people for their special talents. People will be used in reasoning, designing and decision making capacities. They will keep the factory "well oiled". Robotics, CAD/CAM (Computer Aided Design/Computer Aided Manufacturing), ICAM (Integrated Computer Aided Manufacturing) and Artificial Intelligence will be used to do the productive work.

People will maintain the inventory lvels, supplies, services, sales. They will create new products and processes, and act to eliminate production bottlenecks. Much of the dehumanizing aspects of the industrial revolution will be eliminated by the use of "intelligent" machinery.

The transition to a new economy will require sophisticated managers who train and develop their people for future needs. Every revolution brings with it people who fear the future. In 1830, 72% of the U.S. labor force worked on farms. In 1980, one hundred and fifty years later, only 3% of the work force worked on farms. Farming is our most productive area. The farm workers were obviously absorbed into the growing industrial revolution. The United States is currently going through an information revolution.* People working in factories can be retrained to take new factory jobs, or they can be retrained to participate in the new information growth areas.

To ease the problems caused by the transient displacement of blue collar workers, management sensitivity is essential. We can avoid great strife in this country if management accepts the responsibility for retraining its blue collar personnel. If we in management retrain our workers in needed new skills, they will not become a burden to society.

At the end of World War II, West Germany copied our G.I. Bill of Rights. They replaced the word veteran with worker. Major corporations operating in West Germany are required to retrain their workers after a given period of service if the worker requests this retraining. West Germany believes that a society of "bum" workers will become a "bum" society.

As usual, the future is in our hands today. We can, through automation, robotics and artificial intelligence, see a

---

*Megatrends John Naisbitt, Warner Books, 1982.

"golden age" of human intellectual, and artistic expansion or a mechanistic top and bottom civilization. We engineers are the prime movers of this future evolution.

Bibliography

1. "Robots: New Faces in the Production Line," Carl Remick, *Management Review,* May, 1979.

2. "The Blue Collar Robot," Robert Sugarman, *IEEE Spectrum Magazine,* September 1980.

3. "Busy Robots Spur Productivity," Roger Allan, *IEEE Spectrum,* September 1979.

148   BIBLIOGRAPHY

4. "Robots Join the Labor Force," *Business Week,* June 9, 1980.

5. "Robots, Models and Automation," Richard Paul, *IEEE Computer Magazine,* July 1979.

6. "Robots to the Rescue?" Leopold Froechlich, *Datamation Magazine,* January 1981.

7. "The Robot Revolution," *Time Magazine,* December 8, 1980.

8. "Flexible Manufacturing Systems, Digital Controls and the Automatic Factory," S. J. Bailey and Kenneth Pluhar, *Control Engineering,* September 1979.

9. "Scientist Gain Toward Visual Robotics," *Industrial Research and Development,* March 1981.

10. "Robots: A New Force in Continuous Arc Welding," *Manufacturing Engineering,* September 1975.

11. "Developments in Programmable Automation," C. A. Rosen and D. Nitzan, *Manufacturing Engineering,* September 1975.

12. "Flexible Automated Manufacturing Workshop," The National Bureau of Standards, May 30–31, 1979, Hamilton Herman Program Manager.

13. "Cooperative Technology Program Evaluation of 9 Candidate Areas for Industry Participation and Support," Sept. 15, 1979, Industrial Research Institute, Hamilton Herman, Program Manager.

14. "Robot Systems," James S. Albus and John M. Evans, Jr., *Scientific American,* February 1976.

15. *Robotics in Practice,* Joseph F. Engelberger, AMACOM, 1980.

16. *How to Build a Computer Controlled Robot,* Tod Loofbourrow, Haydn Book Company, 1978.

17a. "Sensors for Computer Controlled Mechanical Assembly," S. S. M. Wang and P. M. Will, IBM Research Paper #RC 5918.

17b. *Industrial Robots,* Vols. I and II, William R. Tanner, Society of Manufacturing Engineers, 1979.

150   BIBLIOGRAPHY

18. "Robotics: The 3M Approach," *Robotics Today,* Fall 1980.

19. "Exploratory Research in Industrial Modular Assembly," The Charles Stark Draper Laboratory, November, 1974.

20. "Factory Automation," A Sampson Report, Quantum Science Corporation, 1975.

21. "The U. S. Industrial Robot Market," Frost and Sullivan, June 1974.

22. "Industrial Robots," Stamford Research Institute, 1971.

23. "The Business Case For Robotics," Maurice Zeldman, IEEE, Intercon 75 Conference.

24. "Industrial Robots, An Opportunity," Stamford Research Institute, May 1972.

25. "A High Performance AC Position Servo Using a DC Motor," P.M. Will and M. Zeldman, IEEE, 1965.

26. T. Goto et al, "Precise Insert Operation by Tactile Control Robot," *Industrial Robot,* 1974.

27. "Robot Automation — Today's Status," William E. Uhde, presented at the 1978 International Engineering Conference, May 8–12, 1978.

28. "The Justification of an Industrial Robot," Timothy Bublick, presented at the finishing 1977 Conference and Exposition, October 18–20, 1977.

29. "The Robots are Here!", reprinted from Production's Manufacturing Planbook, 1977.

30. "Let the Robot Do it," Robert B. Aronson, reprinted from the November 27, 1975 issue of *Machine Design.*

31. "Robots Reduce Exposure to Some Industrial Hazards," Norman M. Heroux, and George Munson, Jr. Reprinted from pp. 82–88 of July, 1974, *National Safety News,* A National Safety Council Publication (Vol. 110, No. 7).

32. "Robot-Loaded Stamping Presses Keep Pace with Production," Brian D. Wakefield; reprinted from January, 1977, *Production.*

33. "A Systems Approach to Robot Use in Die Casting," W. M. Goldhammer, presented at the Eighth International Symposium on Industrial Robots, May 5–June 1, 1978, copyright International Fluidics Services Ltd., Kempston, Bedford, England.

34. "Robots Thrive in Tough Environments," reprinted from July, 1976, *Foundry Management and Technology*.

35. "Foundaries, Robots, and Productivity," George Munson; presented at the Eighth International Symposium on Industrial Robots, May 5–June 1, 1978, copyright, International Fluidics Services Ltd., Kempston, Bedford, England.

36. "One Big Step For 'Assembly in the Sky,'" David A. Van Cleave, reprinted from November 28, 1977, *Iron Age*.

37. "Digital Technology Enables Robots to 'See'", Sydney F. Shapiro, copyright 1978, Computer Design Publishing Corporation. *Computer Design*.

38. "Brains, Behavior, & Robotics," James S. Albus, Byte Books, Subsidiary of McGraw Hill, N.J. 1981.

39. "Robot Manipulators," Richard P. Paul, The MIT Press, Cambridge, Mass., 1981.

40. "Megatrends," John Naisbitt, Warner Books, 1982.

41. "The Scientific Relevance of Robotics," Allen Newel, AI Magazine, Winter 1980–81.

1984 List of Robot Manufacturers

This listing shows 90 robot manufacturers and distributors that sell robots in the U.S. The current growth of robot manufacturers and distributors has far outpaced the current market need. In time, many of the current manufacturers will leave the field and several of the remaining manufacturers will share the major market.

The companies that actually produce robots make some 200 different robot styles. This listing can be used to form an idea of each manufacturer's or distributor's product line, an idea of their price, and some indication of special features. The table therefore includes each company's address, their phone number, and the following data:

- Robot Type: This listing considers the coordinate system.

- Power Systems: Hydraulic, electric or pneumatic. Multiple listings generally refer to several families of robots.

- Primary Applications: This listing discloses the Robot Companies targeted market. For some of the newer companies this listing may not represent actual installations. If you find an application of specific interest, we suggest a telephone call to obtain an up-to-date catalogue and information about the company's experience.

- Average Price: This listing is for the companies' full range of products. In general, the units over $35,000 are servo systems, and units about $15,000 are pick and place. There are exceptions to this rule, but it generally holds. Money is listed in thousand dollar figures. Due to intense pressure, caused by the large number of companies in the field, these figures can only be regarded as an approximation. They will obviously fluctuate with market share, strategy, and the pressure to be cost competitive in a given market segment.

- Sensory Ability: This listing describes the feedback capabilities and the degree of feedback sophistication.

- Comments: Generally covers special capabilities or limitations of the robot manufacturers line.

When data was not specified or available at the time of publication, we listed "n" for not specified by the company. A telephone call to the manufacturer can generally get the necessary data.

1984 U.S. MANUFACTURERS AND DISTRIBUTORS OF INDUSTRIAL ROBOTS

MANUFACTURER	ROBOT TYPE*	POWER SYSTEM	PRIMARY INDUSTRY APPLICATIONS	AVERAGE PRICE $ × 1000 (USA)	SENSORY ABILITY*	COMMENTS
Accumatic Machinery Corp. 3537 Hill Avenue Toledo, OH 43607 (419) 893-2979	A	Hydraulic	Welding, load and unload presses and machinery	74	N	6 axis robots, payloads of 35 lb and 75 lb Used in die casting, forging, plastics, and machine tool industries
Acrobe Positioning Systems, Inc. 3219 Doolittle Drive Northbrook, IL 60062 (312) 273-4302	A	Electric & Pneumatic	Light duty assembly and parts handling	30	F, T	Payload 15 lb, up to 10 axis control is possible
Admiral Equipment Co. Subsidiary of the Upjohn Co. 305 West North Street Akron, OH 44303 (216) 253-1353	A	N	Assembly and parts transfer	N	N	Automotive car body sealer was demonstrated at the Robotics VI exhibit

Company	Code	Drive	Application	No.	Codes*	Description
Advanced Robotics Corporation Route 79 Newark Industrial Park Bldg. B Hebron, OH 43025 (614) 929-1065	CA	Electric	Welding	79 to 142 Depends on robot selected	F, P, T, V PR	3 sizes of robot are offered; Turnkey welding capability is the primary market
Air Technical Industries 7501 Clover Avenue Mentor, OH 44060 (216) 951-5191	A, C, S styles available	Electric & hydraulic styles available	Machine loading, light and heavy assembly	10 to 175 Depends on robot selected	F, P, T, V	24 different robot styles are available with payloads that range from 10 lb. to 2000 lb.
American-Monarch Machine Co., Inc. 503 North Niles Metamora, IL 61548 (309) 367-2325	C	Hydraulic	Pick and place parts transfer.	12	N	Simple transfer mechanisms. 3 styles offered with 10 lb., 20 lb., and 32 lb. payloads

*Codes listed on page 187.

159

U.S. MANUFACTURERS AND DISTRIBUTORS OF INDUSTRIAL ROBOTS (CONTINUED)

MANUFACTURER	ROBOT TYPE*	POWER SYSTEM	PRIMARY INDUSTRY APPLICATIONS	AVERAGE PRICE $ ×1000 (USA)	SENSORY ABILITY*	COMMENTS
Anorad 110 Oser Avenue Hauppauge, NY 11788 (516) 231-1990	CA	Electric	Light electronic parts assembly and parts transfer	N	P, F, V, PR	Milling machine like structure. Ultra high repeatability for small strokes (.0001″/inch of stroke)
Armax Robotics Inc. 38700 Grand River Ave. Farmington Hills, MI 48018 (313) 478-9330	C	Hydraulic	Die casting, machine loading and unloading, contouring, welding	N	P, F	One of the older robot companies (Originally the Planet Robot)
Asea, Inc. 1176 E. Big Beaver Road Troy, MI 48084 (313) 528-3630	A	Electric	All applications, including parts transfer, assembly, glueing, inspection, & cleaning	20 to 120 depends on robot selected	P, F, V, RF	3 robot styles are offered: 6.6, 30 and 66 lb. payloads. A soft servo option which can be used for polishing is offered.

Company	Styles	Type	Application	%	Codes*	Description
Automatix, Inc. 217 Middlesex Burlington, MA 01803 (617) 667-7900	CA, A styles available	Electric	Assembly and welding with specialized vision systems	85 to 95 depends on robot selected	F, P, T, V, PR	Turn key welding installations Artificially intelligent vision systems, two styles are offered
The Bendix Corporation Bendix Robotics Division 21238 Bridge Street Southfield, MI 48037 (313) 352-7700	S, A styles available	Electrical	Assembly, transfer and machine center applications	40 to 95 depends on robot selected	F, V, PR	5 styles offered with payload capacity ranging from 7 to 150 lb.
Binks Manufacturing Company 9201 West Belmont Avenue Franklin Park, IL 60131 (312) 671-3000	A	Hydraulic	Spray painting	75	P, T, V	Multi program capability. 8 different paint routines are standard.
Cincinnati Milacron 213 S. West St. Lebanon, OH 45036 (513) 932-4400	A	Hydraulic & Electric styles offered	Robot integration into machine cells. Assembly, loading and unloading, paint, drilling, welding, inspection	65 to 90 depends on robot selected	P, T, V, PR	14 lb. to 225 lb. payloads. Special wrist design. 5 sizes are offered. Computer console permits tracking a moving line capa-

*Codes listed on page 187.

U.S. MANUFACTURERS AND DISTRIBUTORS OF INDUSTRIAL ROBOTS (CONTINUED)

MANUFACTURER	ROBOT TYPE*	POWER SYSTEM	PRIMARY INDUSTRY APPLICATIONS	AVERAGE PRICE $ x 1000 (USA)	SENSORY ABILITY*	COMMENTS
						bility. Company offers Turn key machine center systems
Comet Welding Systems 900 Nicholas Blvd. Elk Grove, IL 60007 (312) 956-0126	CA & C styles available	Electric	Welding	70 to 135 Based on robot style selected	T	Two styles are offered with through the arc sensing capability. Weld quality can be automated
Commercial Cam Division of Emerson Electric 1444 South Wolf Road Wheeling, IL 60090 (312) 459-5200	C	Electric or Hydraulic	Parts handling	5	N	Mechanical cam driven pick & place robot
Control Automation, Inc. P.O. Box 2304 Princeton, NJ 08540 (609) 799-6026	CA	Electric	Parts assembly, parts handling and inspection	37 to 69 depends on robot selected	F, P, T, V, FR	Servo driven robots with vision. Two styles are available with 5

Company	Code	Drive	Application	Price ($1000)	Codes	Description
Copperweld Robotics, Inc. A Copperweld Enterprise 1401 E. 14 Mile Road Troy, MI 48084 (313) 585-5972	C	Electric	Assembly, loading and unloading, for die casting, machine tools & electronics	15 to 45 depends on robot selected	P, V, PR	4 styles of pick and place robots with payloads from 2.5 to 25 lbs are available and 10 lb payloads.
Cybotech Corp. 3939 W. 56th Street Indianapolis, IN 46254 (317) 298-5136	A, C, CA, S styles available	Hydraulic & Electric	Large and small parts assembly. Parts transfer, spot welding, continuous arc welding, cutting, routing, drilling, inspection, and machine cell center.	80 to 300 depends on robot selected	F, P, T, V, PR	9 styles of robots are offered with all types of coordinate systems, and load capabilities from 11 lb. to 175 lb. payloads. Programs up to 400,000 pts. on some systems
Cyclomatic Industries Inc. 7520 Convoy court Sandiago, CA 92111 (619) 292-7440	A, CA styles available	Electric	Welding automation	100 to 125 depends on robot selected	N	Specialty welding robot, 2 sizes are available.

*Codes listed on page 187.

U.S. MANUFACTURERS AND DISTRIBUTORS OF INDUSTRIAL ROBOTS (CONTINUED)

MANUFACTURER	ROBOT TYPE*	POWER SYSTEM	PRIMARY INDUSTRY APPLICATIONS	AVERAGE PRICE $ × 1000 (USA)	SENSORY ABILITY*	COMMENTS
The Devilbiss Company Division of Champion Spark Plug 300 Phillips Avenue Toledo, OH 43612 (419) 470-2296	A	Hydraulic	Paint spray, and welding	N	P, T, V	Two specialty robots are offered. The largest paint robot manufacturer in the world. 60 programs and color changes are possible, in the paint spray robot.
Digital Automation Corp. 65 Walnut Street Peabody, MA 01960 (617) 531-9433	CA	Electric	Micro electronic positioning and assembly	12	F, P, T, V, PR	Small motion precise robot. Repeatability of ± 50 millionths of an inch, and a 2 lb. payload.
Everett/Charles Automation Modules Inc. 700 East Harrison Pomona, CA 91767 (714) 625-5571	CA	Electric	Small parts assembly	22	P, T, V, PR	5 lb. payload assembly robot

Company	Style	Power	Application	Axes	Servo	Description
Feedback Inc. 620 Springfield Av. Berkeley Heights, NJ 07922 (201) 464-5181	A, C styles available	Electric	Small parts transfer	2 to 6 depends on robot selected	N	1 to 3 !b. payloads, six styles of robots, non servo pick and place
H. H. Freudenberg Automation Cobden House, Cobden Street Leicester LEI 2LB England (0533) 531517	C	Electric	Parts transfer	6	N	10 lb. pick and place robot
Gallaher Enterprises 2110 Cloverdale Av., 2B. Winston-Salem, NC 27103 (919) 725-8494	CA	Electric	Specialty electronic assembly robot for PC board insertion	N	P, V	5 lb. payload, up to 6 insertions per minute non servo robot.
GCA/PAR Systems 209 Burlington Road Bedford, MA 07130 (612) 484-7261, x227	CA, C, A styles available	Electric	Parts transfer, loading and unloading	N	P, V, F, PR	12 sizes of robot offered. Anamation simulator programmed in PASCAL and assembly language. Payloads vary from 11 lb. to 2500 lb.

*Codes listed on page 187.

U.S. MANUFACTURERS AND DISTRIBUTORS OF INDUSTRIAL ROBOTS (CONTINUED)

MANUFACTURER	ROBOT TYPE*	POWER SYSTEM	PRIMARY INDUSTRY APPLICATIONS	AVERAGE PRICE $ × 1000 (USA)	SENSORY ABILITY*	COMMENTS
General Electric Co. Robotics and Vision Systems Dept. P.O. Box 17500 Orlando, FL 32860-7500 (203) 382-2000	A, CA, C S styles available	Electric & Hydraulic	Parts transfer, inspection, and assembly, welding, and general industry applications	50 to 125 depends on robot selected	P, T, F	14 robot sizes offered. Robots manufactured by Hitachi of Japan, VW of Germany, and DIA of Italy. Payloads range from 6.6 lb. to 1,300 lb.
General Numeric Corporation 390 Kent Avenue Elk Grove Village, IL 60007 (312) 640-1595	CA, C styles available	Electric	Parts transfer and small electronic assembly	N	F, P	Loads from 11 lb. to 110 lb. Memory up to 6000 points with bubble
GMF Robotics Inc. 5600 New King Street Troy, MI 48098 (313) 641-4242	C, A, CA styles available	Electric	Forging, investment casting, parts transfer, Assembly	14 to 75 depends on robot selected	P, T, F, V, PR	15 robot sizes are offered. Payloads range from 22 lb. to 176 lb.

Company		Drive	Application			Description
Graco Robotics, Inc. 12898 Westmore Avenue Livonia, MI 48150 (313) 523-6300	A	Hydraulic	Spray painting	N	T, P	Self diagnostics, specialty paint spray robot.
Hall Automation Ltd. Boughton Road Rugby CV21 1BU England (0788) 2144	A	Electric	Spray painting & finishing and parts transfer	52 to 67 depends on robot selected	PR, T	Over track mounting system for paint spray robot. 2 paint robots and one 65 lb. payload transfer robot.
Haynes & Foranam Unit 1 Springfield Ind. Estate Farsley, Leeds LS28 7UT U.K. (0532) 55207	N	Pneumatic	Injection molding and parts transfer	5 w/o control	N	Pick and place family of robots with payloads starting at 16 lb. 3 and 4 axis versions available
Hirata Ind. Machineries Co. Ltd. 3-9-20 Togoshi, Shinagawa Tokyo, Japan 142 (0963) 44-2266	C	Electric	Flexible manufacturing and parts transfer	20	N	Payloads vary with number of axes. They range from 16 lb. for 2 axes to 4.5 lb for 4 axes
Hitachi America, Ltd. Industrial Components	C, A, CA styles	Electric	Paint spray, assembly, welding,	N	F, P, T, V, PR	4.4 lb. payload capacity to 22 lb.

*Codes listed on page 187.

167

U.S. MANUFACTURERS AND DISTRIBUTORS OF INDUSTRIAL ROBOTS (CONTINUED)

MANUFACTURER	ROBOT TYPE*	POWER SYSTEM	PRIMARY INDUSTRY APPLICATIONS	AVERAGE PRICE $ × 1000 (USA)	SENSORY ABILITY*	COMMENTS
Sales and Service Division 6 Pearl Court Allendale, NJ 07401 (201) 825-8000	available		machine load and unload, and parts transfer			These robots are sold by many American companies. Paint spray and welding robots are specialty areas. 7 sizes are available
Hodges Robotics Int'l. Corp. 3710 N. Grand River Lansing, MI 48906 (517) 323-7427	S	Electric	Parts assembly, spray paint, weld, and small assembly	55	F, P, T, V, PR	Robot loads range from 35–50 lbs. 8000 point memory systems
IBM Corporation PO Box 1328 1000 NW 51st Street Boca Raton, FL 33432 (305) 998-2000 (914) 765-6630 (White Plains, New York office)	CA, A	Electric and Hydraulic	Electronic and small parts assembly. Sensory capabilities. Inspection and light parts fabrication	35 to 100 depends on robot selected	F, P, T, V, PR	Special robotic language "AML". These robots have been used for 5 years internally for computer and typewriter assem-

bly. 13.2 and 16 lb. payloads

Company	Code	Power	Application			Codes	Notes
Ikegai American Corp. 2246 No. Palmer Drive Suite 108 Schaumberg, IL 60195 (312) 397-3970	CA	Electric	Forging, die casting, loading & unloading, parts transfer & inspection	35 to 44	single or double option	F, P, T, PR	11 and 22 lb. payloads. 2 hands possible. Bubble memory, 4K. Feedback gauging system
INA Automation Forge Lane, Minworth, Sutton Coldfield West Midlands, B76 8AP England	C	Pneumatic	Small and intermediate parts transfer, machine tool loading and inspection	7.4		PR, T	Double gripper available for 2 parts transfer 30 lb. payload
Industrial Automates, Inc. 6123 W. Mitchell Street Milwaukee, WI 53214 (414) 327-5656	C	Hydraulic	Die casting, machine tool loading, parts transfer	13		F, P, T, V, PR	1000 steps keyboard input. 10 lb. payload
International Robomation/ Intelligence 6353 El Camino Real Carlsbad, CA 92008 (619) 438-4424	A	Pneumatic	Die casting, forging, investment casting parts transfer	12.5		P, PR	Pneumatic servo robot computer controlled, 50 lb. payload. This robot is powered by

*Codes listed on page 187.

U.S. MANUFACTURERS AND DISTRIBUTORS OF INDUSTRIAL ROBOTS (CONTINUED)

MANUFACTURER	ROBOT TYPE*	POWER SYSTEM	PRIMARY INDUSTRY APPLICATIONS	AVERAGE PRICE $ × 1000 (USA)	SENSORY ABILITY*	COMMENTS
						5 digital air servo motors for its 5 axis unit.
ISI Manufacturing Inc. 31915 Groesbeck Highway Fraser, MI 48026 (313) 294-9500	CA	Hydraulic and Electric	Die casting, welding, metal stamping, parts transfer	75 to 160 based on robot selected	P, T, PR, F	Programmed in basic machine language
Kuka Welding Systems and Robot Corporation 24031 Research Drive Farmington Hills, MI 48018 (313) 977-0100	A	Electrical	Manufacturing, welding	N	T, V, PR	Multiple arms can be synchronized in this 5 axis specialty welding robot. A transfer robot with a 220 lb. payload is also offered
Lamson Corp. P.O. Box 4857 Syracuse, NY 13221 (315) 432-5500	CA	Electric	Palletizing bricks, bags, cartons, bales	40	Speed, T, V, PR	Up to 100 lb. payload in this 2K memory specialty packaging robot.

Company		Power	Applications		Codes	Description
Lynch Machinery Division 2300 Crystal Street P.O. Box 2477 Anderson, IN 46018 (317) 643-6671	CA	Hydraulic	Die casting, parts transfer, load and unload	20 to 100 based on robot selected	F, PT, V, PR	Payloads of 300 to 1000 lb. in this series made to handle heavy parts
Machine Intelligence Corp. 330 Potrero Avenue Sunnyvale, CA 94086 (408) 737-7960	A	Electric	Assembly, parts transfer and vision system part orientation	80	V	Payload of 6.6 lb. in a specialty "smart" robot with vision
Mack Corp. 3695 East Industrial Drive Flagstaff, AZ 86001 (602) 526-1120	CA	Pneumatic Electric	Light duty bench top parts transfer assembly and packaging	8	PR	5 and 10 lb. payload pick and place device
Manca Inc. 165 Carver Ave. Westwood, NJ 07675 (201) 666-4100	CA	Hydraulic Pneumatic	Parts transfer	N	N	Modular robotics concept
Marlo Company, Ltd. 1-34 2 Chome, Ohashi-Cho	A	Hydraulic	Parts transfer, forging, machine	60 to 70 based on	PR and human oper-	Manipulator payload 2200 lb., ro-

*Codes listed on page 187.

171

U.S. MANUFACTURERS AND DISTRIBUTORS OF INDUSTRIAL ROBOTS (CONTINUED)

MANUFACTURER	ROBOT TYPE*	POWER SYSTEM	PRIMARY INDUSTRY APPLICATIONS	AVERAGE PRICE $ × 1000 (USA)	SENSORY ABILITY*	COMMENTS
Nagata-Ku, Kobe, Japan 078-611-2151			tool loading, and manipulator for human assist.	robot size		bot 1320 lb. payload. Specialty heavy duty 4 axis robot
Microbo S. A. Crocelles (NE) 2035 Switzerland (038) 31-57-31	CA	N	Small parts sequential assembly and transfer multi arm robot	100 to 180 based on robot selected	P, V, F, PT, PR	.2 lb. to 2 lb. payloads for electronic assembly and small mechanism assembly
Microbot, Inc. 435-H Ravendale Drive Mountain View, CA 94043 (415) 968-8911	A	Electric	Light part material loaders and instructional devices	2 to 7.5 based on robot selected	N	Three styles, one is a stepper motor microcomputer device useful for learning about robots. Two others are small parts handlers with payloads of 1 lb. and 1½ lb.

Company						
Middlesex General Industries 6 Adele Road Woburn, MA 01801 (617) 935-8870	CA	Electric	Mechanical assembly small parts transfer, electronic assembly	10	PR	2.2 lb payload, 4 axis robot
Mitsubishi Electric Corp. 2-2-3 Marunochi, Chiyoda-Ku Tokyo, Japan (052) 721-2111	CA	Electric	Welding and parts transfer	N	N	20 lb. payload, 6 axis robot
Mobot Corp. 980 Buenes Ave. San Diego, CA 92110 (619) 275-4300	C, CA, S styles available	Electric	Die casting, machine tool loading/unloading, material handling and palletizing	40 to 90 based on robot selected	F, P, T, V, PR	15 different styles of pick & place robot 10 lb. to 600 lb. payload. Standard 4K memory. This company offers a modular approach to the robot user. You buy what you need.
Modern Machine Works Inc. 5355 South Kirkwood Av. Cudahy, WI 53110 (414) 744-5900	A	Hydraulic	Turnkey automatic machine tool loading system	30	PR	30 lb. payload pick and place transfer robot, 2 axis design

*Codes listed on page 187.

U.S. MANUFACTURERS AND DISTRIBUTORS OF INDUSTRIAL ROBOTS (CONTINUED)

MANUFACTURER	ROBOT TYPE*	POWER SYSTEM	PRIMARY INDUSTRY APPLICATIONS	AVERAGE PRICE $ × 1000 (USA)	SENSORY ABILITY*	COMMENTS
Mouldmation, Ltd. Unit 1C, Chase Park Industrial Estate, Chase Terrace WS7 8JQ, England (05436) 2075	CA	Pneumatic	Molding machine loading and unloading	10.7 to 34 based on robot selected	N	11 lb. or 220 lb. payload, 5 or 6 axis pick and place robot
MTS System Corp Box 24012 Minneapolis, MN 55424 (612) 937-4000	A	Hydraulic	Forging, investment casting, machine tool loading/unloading, parts transfer, welding, machining, inspection.	120	F, P, T, V, PR	220 lb. precision path 6 axis hydraulic servo robot
Nordson Corporation Robotics Division P.O. Box 151 555 Jackson Street Amherst, OH 44001 (216) 988-9411	A	Hydraulic	Spray painting	59 to 110 based on robot selected	P, V, PR	35 lb. and 10 lb. payload robot. 32K memory. Special off-line teaching aid.

174

Company	Code	Drive	Applications		Code	Remarks
NOVA Robotics Inc. 265 Prestige Park Road East Hartford, CT 06108 (203) 528-9861	CA	Electric	Small parts loading & unloading, electronic assembly, inspection	65	F, P, T, V, PR	40 lb. payload for 3 axis robot and 15 lb. payload for 6 axis robot
Osaka Transformer/Merric Eng. Inc. 5-1 Minami-Senrioka, Settsu Osaka, 564 Japan (06) 381-7041	C	Electric	Welding (MIG, TIG, plasma)	75	N	2 styles of welding robot are supplied
Pendar Robotics Ltd. Rassau Industrial Estate Ebbw Vale Gwent, S. Wales NP3 5SDUK (0495) 307070	A	Pneumatic	Machine tool loading & unloading, stacking, palletizing, paint spray	30	P, T, V, PR	4.4 lb., 11 lb., and 22 lb. payloads
Pharemme S. A. Les Nouhauts 87370 Saint Sulpice Lauriere France (55) 714411	A	Hydraulic	Machine tool loading/unloading parts transfer, small parts assembly, welding and paint spray	40	N	55 Payload 6 axis robot

*Codes listed on page 187.

175

U.S. MANUFACTURERS AND DISTRIBUTORS OF INDUSTRIAL ROBOTS (CONTINUED)

MANUFACTURER	ROBOT TYPE*	POWER SYSTEM	PRIMARY INDUSTRY APPLICATIONS	AVERAGE PRICE $ × 1000 (USA)	SENSORY ABILITY*	COMMENTS
Pickomatic Systems 37950 Commerce Sterling Heights, MI 48077 (313) 939-9320	Multi-arm index drive mech-anism	Electrical	Mechanical parts handler	51	N	This device is a combination fixed automaton and ro-bot technology
Positec Corp. Rush Lake Road Laurens, IA 50554 (712) 845-4548	C, CA styles available	Hydraulic Electric Pneumatic	Investment, cast-ing, loading and unloading, spray paint, custom built	90	P, T, F, RR	This hybrid robot can carry up to 1250 lb. payload
Prab Robots, Inc. 6007 Sprinkle Road Kalamazoo, MI 49003 (616) 329-0835	C, S styles available	Hydraulic	Machine tool load/unload, part transfer, washing automation, as-sembly, welding, paint spray, in-spection, glass transfer, and pal-letizing.	30 to 150 based on robot selected	P, T, F, V, PR	12 different styles of robot are of-fered with pay-loads from 100 lb. to 2000 lb. up to 7 axes versions available. The SERVO robots are the original Versa-

Company		Applications		Power	Codes*	Description
						tran, the second oldest robot company in the U.S. Prab purchased Versatran in 1979.
Precision Robots, Inc. Carmel Circle Lexington, MA 02173 (617) 862-1124	A	Machine tool loading packaging and assembly	30	Electric Pneumatic	P, F, V	2 lb. payload robot designed for assembly
The R-2000 Corporation 804 Broadway West Long Branch, NJ 07764 (800) 221-0535	N	Plastic industry loading and unloading and parts transfer.	25	N	P, F, V, PR	5 lb. payload, 5 axis pick and place robot
Reis Machines 1426 Davis Road Elgin, IL 60120 (312) 741-9500	A	Die casting, forging, machine tool loading, parts transfer, nuclear waste manipulation, palletizing	65 to 75 based on robot selected	Electric	P, F, PR, T	55 and 110 lb. payload, 6 axis Horizontal anthropomorphic robot

*Codes listed on page 187.

U.S. MANUFACTURERS AND DISTRIBUTORS OF INDUSTRIAL ROBOTS (CONTINUED)

MANUFACTURER	ROBOT TYPE*	POWER SYSTEM	PRIMARY INDUSTRY APPLICATIONS	AVERAGE PRICE $ ×1000 (USA)	SENSORY ABILITY*	COMMENTS
Remek Micro Electronics, Ltd. 35, Barton Road, Water Eaton Industrial Estate Bletchley, Milton Keynes MK2 3HY, UK (0908) 649494	CA	Electric	Machine tool loading, parts transfer, small parts assembly, spot welding, automatic screw-driver	22	P, F, V, PR	Self optimizing ro-bot for minimum cycle time. 22 lb. payload, 3 axis
Resh Industries 3420 W. Capitol Drive Milwaukee, WI 53216 (414) 444-9980	A	Hydraulic Air	Machine tool loading, parts transfer	80	PR, F	Two sizes availa-ble: 50 lb. and 200 lb. payload and in 5 or 6 axis config-urations, pick and place robots

Manufacturer		Drive	Applications		Codes	Comments
Rhino Robots, Inc. PO Box 4010 3402 Mattis Ave. Champaign, IL 61820 (217) 352-8485	A	Electric	Loading machines, and transfer of light parts	3 to 27 depends on robot selected	P, T, PR, V	3 lb., 5 lb. and 15 lb. payloads for six axis, pick and place robots
RoMec Inc. 1000 Tech Park Drive Billerica, MA 01821 (617) 667-8944	C, CA styles available	N	Machine tool loading, small parts assembly, electronic assembly and welding	45	PR, V, T	17 to 55 lb. payloads designed for workstation assembly and parts transfer
Satt Kaufeldt AB Box 32006 Tellusborgsvagen 12611 Stockholm (08) 810100	C	Pneumatic	Die casting, forging machine tool loading, parts transfer		PR	3 models are offered in 11 lb., 55 lb. and 110 lb. payloads for pick and place type robots. Double sided grippers for parts handling from each side

*Codes listed on page 187.

U.S. MANUFACTURERS AND DISTRIBUTORS OF INDUSTRIAL ROBOTS (CONTINUED)

MANUFACTURER	ROBOT TYPE*	POWER SYSTEM	PRIMARY INDUSTRY APPLICATIONS	AVERAGE PRICE $ × 1000 (USA)	SENSORY ABILITY*	COMMENTS
Schrader Bellows Division Scovill Manufacturing 200 W. Exchange Street Akron, OH 44309 (919) 556-4031	C	Pneumatic	Die casting, load and unload machines, small parts assembly, forging plastic parts	10	P, PR	5 lb. payload 5 axis. Pick and place robots
Seiko Instruments USA, Inc. 2990 W. Lomita Blvd. Torrance, CA 90505 (213) 530-8777	CA, C styles available	Pneumatic & Electric	Small parts assembly, electronic assembly, inspection, machine loading and unloading, palletizing and packaging	5 to 29 depends on robot selected	P, T, V, PR	1.1 lb., 2.2 lb. and 11 lb. payloads for precision DC servo robots and eight sizes of pick and place robots with payloads from 1.6 lb. to 8.8 lb. small motion, high precision

180

Company	Code	Power	Application	Value	Sensing	Notes
Shin Meiwa Industry Co., Ltd. Shin Ohtemachi Bldg. 6-1 Ohtemachi 2 Chome Chiyoda-Ku, Tokyo Japan (312) 346-9339 (in Chicago)	CA, A styles available	Electric	Welding & turn key welding installations for large and intermediate parts	70 to 163 depends on robot selected	Through the arc sensing	Three sizes of welding robot are offered with a special feature of accurate welds on parts that are placed approximately
Shoku Corporation/Eng. Dept. 1010 Minorudai Matsudo City, Chiba Pref., Japan (0473) 64-1211	CA	Pneumatic	Die casting, machine loading & unloading	14 to 22 depends on robot selected	N	6.6 lb. and 11 lb. payloads are available
Sigma Sales Inc. 6505 Serrano Ave. Suite C Anaheim, CA 92807 (714) 974-0166	A	Electrical	Small parts loading and assembly	5.5	P, T, F, V, PR	This device is operated from a micro computer keyboard teach mode with basic or FORTRAN language

*Codes listed on page 187.

181

U.S. MANUFACTURERS AND DISTRIBUTORS OF INDUSTRIAL ROBOTS (CONTINUED)

MANUFACTURER	ROBOT TYPE*	POWER SYSTEM	PRIMARY INDUSTRY APPLICATIONS	AVERAGE PRICE $ ×1000 (USA)	SENSORY ABILITY*	COMMENTS
Sormel S. A. 6 Rue Becquerel Z. I. Chateaufarine, B.P. 1565 25009 Besancon Cedex, France (617) 890-9100 (Mass.)	CA	Electric	Automatic flexible assembly, parts loading, parts transfer, automatic tool changing, machine tool loading	13 to 21 depends on robot selected	N	3.3 lb. and 77 lb. payload robots for machine cells
Spadone Machine Company, Inc. 507 Westport Avenue Norwalk, CT 06856 (203) 846-1677	CA, C styles available	Pneumatic	Small parts transfer and machine loading	3.5 to 4 depends on robot selected	PR	Payload of 5 lb. for cylindrical and rectangular coordinate pick and place robots
Sterling Detroit Company 261 E. Goldengate Avenue Detroit, MI 48203 (313) 366-3500, ext. 3	A and special limited motions robot	Hydraulic	Die casting, loading and unloading, parts transfer, inspection and machine centers	16 to 25 depends on robot selected	F, PR, P, V	200 to 700 lb. payloads in 3 sizes of robot are offered

Company	Style	Drive	Applications		Codes*	Features
Thermwood Robotics 1910 Firman Drive Suite 110 Richardson, TX 75081 (812) 937-4476	A, CA styles available	Hydraulic	Woodworking, parts transfer, spray painting, and specialty areas	33 to 65 depends on robot selected	P, T, PR	18 to 25 lb. payloads for 2 sizes of robot. Several versions of robot motion tables are used for woodworking, routing and finishing. Specialty paint and weld robots are also in the product line.
Tokico America, Inc. 3555 W. Lomita Blvd. Suite E Torrance, CA 90505 (213) 534-3300	A	N	Paint spray	N	P, T, PR, V	Micro computer operation capable of being synchronized with conveyor systems
Design Technology Corp. 53R Second Avenue. Burlington, MA 01803 (617) 229-6670	A	Electric	Electronic assembly, screwdriver, wiring, soldering, parts transfer, machine loading, inspection and palletizing	N	P, T, PR, V, T	5.5 lb, 22 lb, and 55 lb. payload robots are offered. These come in 3, 4, and 6 axis versions. Features include synchronization of a pair of

*Codes listed on page 187.

183

U.S. MANUFACTURERS AND DISTRIBUTORS OF INDUSTRIAL ROBOTS (CONTINUED)

MANUFACTURER	ROBOT TYPE*	POWER SYSTEM	PRIMARY INDUSTRY APPLICATIONS	AVERAGE PRICE $ × 1000 (USA)	SENSORY ABILITY*	COMMENTS
						robots in assembly activities. A.C. Servo Drive System
Unimation, Inc.* Shelter Rock Lane Danbury, CT 06810 (203) 744-1800 *division of Westinghouse	S, A styles available	Electrical Hydraulic	Die casting, forging, machine loading, parts transfer, small parts assembly, electronics assembly, machining, welding inspection, drilling, and palletizing	30 to 80 depends on robot selected	F, P, T, V, PR	14 styles of robots are offered with payloads from 10 to 450 lb. This is the original robot company (the industry's grandfather). Features include a special robot language called VAL. And specialty robots for welding.

184

Company		Type	Application		Configuration	Description
United States Robots 650 Park Ave. King of Prussia, PA 19406 (215) 768-9210 Conshohocken, PA 19428 (215) 825-8550	S	Electrical	Machine tool loading, electronic and small parts assembly, transfer and inspection	37	P, PR	5 lb. payload robot with remote diagnosis over telephone lines, 32K memory
United Technologies Steelweld Robotic Systems 5200 Auto Club Drive Dearborn, MI 48126 (313) 593-9600	Special indexing unit and A, CA, robot styles available	Electrical	Small parts assembly, electronics assy., loading and unloading of machine tools, arc welding, flexible automation, turn key welding capability	N 47 to 112 depends on robot selected	V, P, F, PR, T, and voice	7 robot styles offered with payloads of 5.5 lb., 11 lb., 33 lb., 44 lb. and 132 lb. Specialties include welding with arc sensing capability and voice actuation
Westinghouse Electric Corp. * Industry Automation Division 400 Media Drive	S, CA, C, A styles available	Electrical	Small, intermediate and large parts for all industry applications. Application	N	F, P, T, V, PR and voice	7 robot styles are featured. Small assembly robots for assembly to large welding robots.

---

*Codes listed on page 187.

185

U.S. MANUFACTURERS AND DISTRIBUTORS OF INDUSTRIAL ROBOTS (CONTINUED)

MANUFACTURER	ROBOT TYPE*	POWER SYSTEM	PRIMARY INDUSTRY APPLICATIONS	AVERAGE PRICE $ × 1000 (USA)	SENSORY ABILITY*	COMMENTS
Pittsburgh, PA 15205 (412) 244-2000 *now merged with Unimation						Payloads from 22 lb. to 300 lb. Sophistication goes from standard to full sensory robots. Turnkey manufacturing and welding systems. Productivity assessments
Wexford Robotics 2118 Queen Regina, Saskatchewan Canada S4T4C3 (306) 522-7429	S	Electrical	Investment casting, small parts assembly, inspection, cleaning	100	F, P, T, PR	This robot has a 50 lb. load carrying capacity. Assists in the design and development of robots and peripherals.
Wickman Automation Ltd.	C		Die casting, forging	30	T, PR	11 lb. payload

186

Printed and bound by CPI Group (UK) Ltd, Croydon, CR0 4YY

28/10/2024

01780048-0001

Company	Type	Payload	Applications	Sensors	Notes
Herald Way, Brandon Road Binley, Coventry, England CV3 2NY U.S. Branch 225 Industrial Drive Warwick, RI 02881			ing, machine tool loading, small parts assembly, finishing, and inspection		
Yaskawa Electric America, Inc. Motoman Division 305 Era Drive Northbrook, IL 60062 (312) 564-0770	Electrical	51 to 86 depends on robot selected	Heavy machine loading and small parts and electronic assembly, and inspection, sealing and glueing. Specialty welding robots	F, P, T, PR	A	5 styles available with payloads of 6.6 lb., 22 lb., 66 lb. A special feature is the electrical system. It is a PWM AC drive. (Pulse Width Modulated AC Drive System)

*Codes

A = Anthropomorphic
C = Cylindrical
CA = Cartesian
S = Spherical
N = Not specified

F = Force Feedback
P = Parts Detection
T = Tracking Sensors
V = Vision
PR = Proximity Detectors

187

Index

A

S

T

U

V

W